T0092662

Factfulness Sustainability

Thomas Unnerstall

Factfulness Sustainability

What you should know about Ecological Crises and Resource Consumption

Thomas Unnerstall
Gernsheim, Hessen, Germany

ISBN 978-3-662-65557-3 ISBN 978-3-662-65558-0 (eBook)
https://doi.org/10.1007/978-3-662-65558-0

© Springer-Verlag GmbH Germany, part of Springer Nature 2022
The translation was done with the help of artificial intelligence (machine translation by the service DeepL.com). A subsequent human revision was done primarily in terms of content.
This work is subject to copyright. All rights are reserved by the Publisher, whether the whole or part of the material is concerned, specifically the rights of translation, reprinting, reuse of illustrations, recitation, broadcasting, reproduction on microfilms or in any other physical way, and transmission or information storage and retrieval, electronic adaptation, computer software, or by similar or dissimilar methodology now known or hereafter developed.
The use of general descriptive names, registered names, trademarks, service marks, etc. in this publication does not imply, even in the absence of a specific statement, that such names are exempt from the relevant protective laws and regulations and therefore free for general use.
The publisher, the authors, and the editors are safe to assume that the advice and information in this book are believed to be true and accurate at the date of publication. Neither the publisher nor the authors or the editors give a warranty, expressed or implied, with respect to the material contained herein or for any errors or omissions that may have been made. The publisher remains neutral with regard to jurisdictional claims in published maps and institutional affiliations.

This Springer imprint is published by the registered company Springer-Verlag GmbH, DE, part of Springer Nature.
The registered company address is: Heidelberger Platz 3, 14197 Berlin, Germany

Preface to the English Edition

This book was first published in German a year ago, in March 2021. From the start, however, it was written and intended for an international audience, and so I am very glad that it is now available in English.

As two years have passed since I finished the original manuscript, I have updated the figures and conclusions, where applicable, with the latest available information. It was quite interesting to see that all major ecological trends described in this book are corroborated by these new data.

With this presentation of the essential facts about the main sustainability issues of our time, for the world as well as for the USA and the EU, I hope to achieve two goals:

- Correct the distorted and gloomy picture constantly conveyed by most media on these issues – the future here is much brighter than you might think.
- Convince the reader that economic development, higher living standards and market economies are not the problem (as often suggested), but rather part of the solution – also, and in particular, with respect to changing our energy systems to CO_2-free energies in the next decades.

If my work contributes just a tiny bit to a more rational, fact-based view of the world, I will be greatly rewarded.

It is a pleasure to thank Springer-Verlag and especially Margit Maly for making this English edition possible.

Gernsheim
Hessen, Germany
April 2022

Thomas Unnerstall

Preface to the German Edition

The story of this book begins in a train station-bookstore in November 2018. I just want to take a quick look to see if there are any new books on energy transition/climate protection (the topic I have mainly written about up to now), but then a title catches my eye: "Mankind is abolishing itself." "What a statement" I think to myself and take a closer look. It's by Harald Lesch, a well-known scientist, and it's about humanity's consumption of resources, about global ecological crises, about climate change. The cover is emblazoned with a red sticker: SPIEGEL Bestseller. Enough arguments to buy the book.

A few hours later, I'm sitting in the train, reading sentences like "Excessive resource consumption in Germany," "We're fishing the seas empty," "The planet is dying of thirst," "The rainforest is burning" – and I start to have doubts. Why do I doubt? First, the author's arguments for these far-reaching statements seem to me to be incomplete and unsystematic; second, I have recently read another book ("Dare the future" by M. Horx) which is much more positive about the future relationship between man and nature. "Now, who is right?" I mutter to myself. Over the next few weeks, the decision matures in me to actually get to the bottom of this question: What is the real state of our planet, how serious are environmental damage and resource consumption by humans – judged as systematically as possible and strictly on the basis of all available facts? Do we need the "radical change" that H. Lesch calls for in his book and that many other authors call for?

Of course, I did not realize at the time what this undertaking actually entailed, that it would take up the next two years of my life. But I have not regretted the decision. It was always exciting to pin down and understand the relevant data and contexts behind the questions at stake: "How much iron has mankind consumed so far, and what percentage of total iron stocks is that?"; "To what degree is future food production really threatened by soil erosion?"; "How has species extinction developed in recent decades?"; "What does the 'ecological footprint' of mankind actually mean?"; "Would it be sustainable if every human being consumed as much energy as we do in the West?"; and many more. And there were many surprises: Often, when all the facts were on the table in front of me after weeks or months of research, I found that the data yielded a completely different picture than the one given or at least suggested in most media (and partly in H. Lesch's book) about the respective question.

The book you are holding in your hand (real or via screen) is the result of this path. I am looking forward to your judgement.

* * *

My first thanks regarding this book must go to the inventors of the Internet. It would have been practically impossible to write such a book 30 years ago: The necessary data, statistics and research results (if available at all) would have been buried in individual libraries around the globe, almost impossible to find … a hopeless undertaking. Today, in principle, any person – completely independent of social circumstances, of his or her culture and religion, of origin and gender – with Internet access and enough time can write such a book; all relevant databases, reports and scientific articles are available online. (We can be curious what effects this fantastic availability and thus democratization of almost the entire knowledge of mankind, unimaginable only a few decades ago, might have in the next decades.)

I would like to express my sincere thanks to my friends Ulrich Parlitz, Manuel Rink, Annette Schild and Corinna Spott, who critically read the manuscript and greatly enriched it with numerous hints, suggestions and objections. Ulrich Dieckert, Jessica Korth and Harald Notter also helped me in developing my thoughts.

Finally, a big thank you is due to Barbara Lühker and Margit Maly for the trustful and constructive cooperation with Springer-Verlag.

Gernsheim
Hessen, Germany
September 2020

Thomas Unnerstall

Contents

1 Introduction

On July 28th, 2022, the time had come once again: "Earth Overshoot Day" 2022 was proclaimed. Every year, "Earth Overshoot Day" is the day on which humanity – according to the generally accepted concept of the "ecological footprint" – has used up the Earth's resources allotted to it for the year. From that day on, the statement goes, humanity lives ecologically "on credit," plundering the planet in an unsustainable way. Or, to put it the other way around: humanity would need 1.8 Earths to maintain its current level of consumption in the long term, and the trend is rising. The situation is said to be even worse with regard to the Western way of life: by the same token, the USA would need five Earths, the EU countries still three.

Figure 1.1, which illustrates this statement, has been printed countless times in the world's media: it shows that, already since 1970, there has been this "overshoot" of human consumption over the sustainably available resources of planet Earth. So far, so clear.

Really?

Take a closer look at this figure. The picture is clearly dominated by the red area, which stands for CO_2 emissions. Quite obviously, the whole problem of greenhouse gases and climate change plays a major, even a decisive role in the aforementioned statements. Taking *this* ecological

© Springer-Verlag GmbH Germany, part of Springer Nature 2022
T. Unnerstall, *Factfulness Sustainability*,
https://doi.org/10.1007/978-3-662-65558-0_1

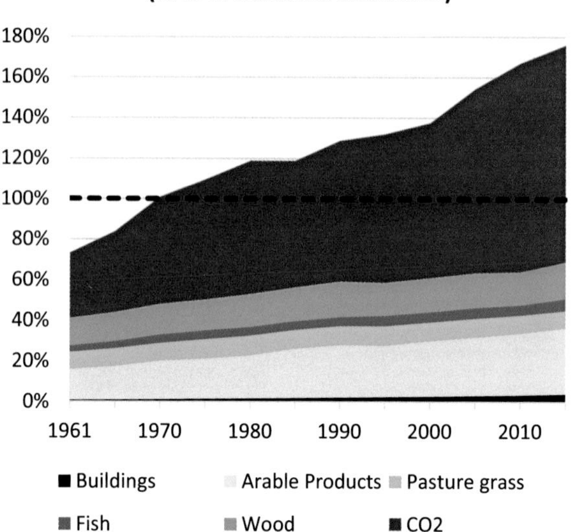

Fig. 1.1 Humanity's "ecological footprint" as a percentage of the Earth's available renewable resources (the so-called "biocapacity"). The more precise definitions are explained in Chap. 11. (Sources: Global Footprint Network, own calculations)

issue out of the picture, we look at Fig. 1.2, the so-called "non-energy" ecological footprint of mankind.[1]

This figure now says something completely different: The resource consumption of mankind with regard to arable land, pasture land, fishing grounds and forests is actually fine: we consume less than 70% of the available resources of the planet. What's more, even in 2050 – when there will be almost 10 billion people – humanity will most likely not consume more than 80% of these available natural resources[2]; except for the issue of CO_2 emissions.

[1] I adopt this notion from Randers (2012).

[2] See Chap. 11 for the explanation.

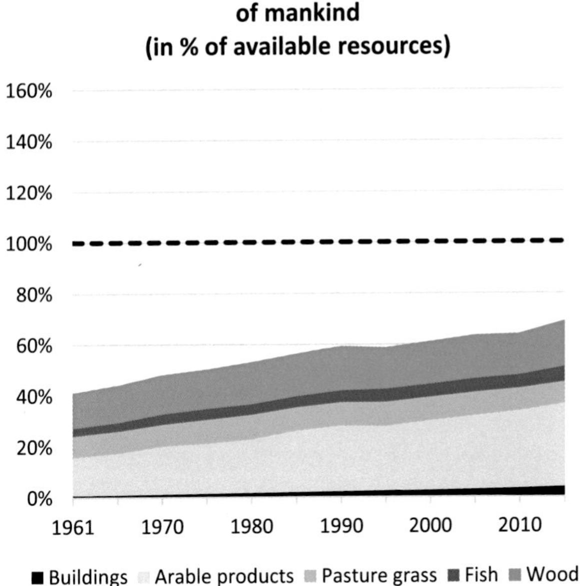

Fig. 1.2 The "ecological footprint" of humanity, as a percentage of the Earth's available renewable resources, excluding the carbon footprint. The more detailed definitions are explained in Chap. 11. (Sources: Global Footprint Network, own calculations)

This question "Really?," that is, a second and sobre look at the numerous statistics on today's global ecological hotspots – often referred to as "ecological crises" – is the **core of this book**. Just as it is worth questioning the headline

"Humanity consumes 1.8 Earths",
it is worth taking a closer look at the equally common headlines (as a selection):
"Rainforests are disappearing at a breathtaking rate"

"Humanity is causing the sixth mass extinction in Earth's history"
"By 2050, there will be more plastic than fish in the world's oceans"
"In the next decades, drinking water will become scarce"
"The earth's raw materials are running out"

That is precisely what I will do in this book, often with surprising results. I will discuss the essential ecological issues that have been on the agenda in the last decades at the relevant global institutions – organizations of the UN; internationally active NGOs like Greenpeace, WWF, IUCN; international think tanks and others – as well as in public debates and in the media.

In doing so, I will not present any new data or scientific findings on the ecological state of the Earth. The methodological approach is rather to look at the existing data and graphics more closely and neutrally than is usually done, and to put them into the right context. In this way, piece by piece, the book sketches a new picture of the impact of man on nature, of the Earth's available resources for the world economy and world consumption, and of the development of ecological hotspots up to 2050. It is a picture rather different from the one we all regularly get from the media.

The **second intention** of this book also suggests itself from the comparison of Figs. 1.1 and 1.2. CO_2 emissions and the global warming they cause clearly dominate humanity's ecological footprint; and climate change has also dominated the socio-political discussions about (ecological) sustainability in our time for a number of years now. Indeed, there is rare unanimity – at least in terms of understanding and declarations of political goals – among almost all the world's governments: climate change is an absolutely serious, very urgent challenge for humanity.

I am convinced, and assume in this book, that this assessment is true. Against this background, the book explores the question of how serious and how urgent, *compared to climate change*, the other sustainability issues really are – resource consumption, scarcity of raw materials, species extinction, rainforest deforestation, plastic waste. Do they also threaten our livelihoods to a similar extent?

To put it more boldly: Are all of these problems more or less equally bad, and are we therefore doomed, heading toward ecological catastrophe one way or another in the course of the twenty-first century? Or are there essential differences and, consequently, essential priorities for action?

The book provides a clear answer to this question.

In the debates about the impact of human life, economy and consumption on ecosystems and thus also on our own livelihoods, the focus is often on **the West**. The Western countries – primarily the USA and the European Union (EU) – are often said to be the main culprits for the irresponsible exploitation of nature. The capitalist, profit-oriented economic system, the excessive, ever-increasing consumption, and in the end the entire Western lifestyle – these factors are alleged to have led to the ecological crises of the present and to be incompatible with a sustainable way of life that respects ecosystems, animals and plants. This then results in the claim that a "great transformation" is indispensable[3] – in other words, a fundamental reform of our way of doing business, consuming and living.

Similar to the above-mentioned headlines about rainforests, species extinction, plastic waste, and raw material shortages, this narrative has become engrained in the consciousness worldwide – and especially in the minds of many of those who follow current events closely, read books like

[3] Cf. Schneidewind (2018).

this one, and are concerned about the future of our children and grandchildren.

The **third intention of** the book is to ask, again, "Is this really true?" Do the relevant scientific findings, data, and statistics really allow us to draw these conclusions? In order to answer this question, for every topic I will present not only the relevant global data, but also the respective figures related to the West, more specifically to the U.S. and the EU combined.[4]

The **structure of** this book is different from usual: I put the main conclusions at the beginning.

Thus, Part I consists of three theses, which I substantiate on the basis of a summary of the following parts II to IV. They answer the following questions:

- What are the priorities for the global ecological problems?
- Is the Western economic system and way of life really "unsustainable"?
- What is the record so far with respect to the political and social efforts to improve the environment?

Part II then deals with the biological foundations of human existence on earth. Here, the focus is on the question of whether enough arable land, food and drinking water are available for a growing world population.

In Part III, the essential building blocks for human economic activity on the planet are scrutinized: energy and raw materials. Global consumption of oil, gas and coal, as well as iron, copper, aluminum, phosphorus and others, has exploded in the last 50 years. Will we soon reach "the limits of growth"?

[4] Of course, it would have been possible to present the data for the USA and for the EU separately. I have refrained from doing so in order not to add to the already considerable density of information. All data on the EU refer to the EU of 28 countries, i.e., including Great Britain.

Finally, Part IV is devoted to the central ecological hotspots (with the exception of climate change) that have been in the focus of world attention for the past 20–30 years:

- humanity's ecological footprint,
- biodiversity and species extinction,
- forest loss/deforestation of rainforests,
- plastic waste in the world's oceans,
- dead zones and P/N cycle,
- pollution by harmful substances.

For each of these issues, I present the historical development over the past 50–60 years, the status quo and the expected development in the next decades. Each chapter here concludes with an assessment of the long term impacts on our planet's ecosystems and on the livelihoods of future generations, often also addressing the effects of climate change.

The whole book solely rests on information, data, and research available online. Most of the chapters are based on reports and/or databases of international organizations: FAO, World Bank, UNEP, IUCN, Global Footprint Network, World Wildlife Fund, World Steel Association, etc. From time to time, I have also made use of websites that provide the public with information that has already been prepared, e.g. www.worldometers.info and www.ourworldindata.org. Not infrequently, it was also necessary to include the relevant research literature in order to complete the overall picture with current findings or additional contexts. The numerous illustrations are based throughout on the figures and data from the sources indicated in each case; however, in most cases these have been newly compiled and categorized ("own calculations").

With this book, my aim is provide an overview of the status, perspectives and importance of the central ecological hotspots of our time in an easily accessible form. The structure, language and style of the book have been chosen to meet this goal:

- Each chapter focuses on the (in my view) most important aspects and contexts of the topic in question. To do this, it was necessary to simplify complex issues considerably and to dispense with many differentiations that might be important or even indispensable from the point of view of a particular expert.
- In many cases, more detailed figures or additional aspects have been relegated to the footnotes. You can therefore decide for yourself to what extent you wish to take in this additional information.
- All figures are generously rounded to promote a focus on the essentials and quick comparisons. The appropriate markings ("circa," "roughly," "around") have often been omitted.
- In the so-called "excursions" I have deepened the topic of a chapter by means of an example or an aspect which seemed to me to be particularly concise or impressive. However, they are not necessary for understanding the main text.

A well-founded overview of ecological sustainability in this first half of the twenty-first century on about 250 pages – in view of the abundance of relevant topics, this is only possible if one restricts oneself to the essentials, to overarching statements and assessments. Conversely, this means that I was forced to leave many important aspects unmentioned: particular grievances, local developments against the general trend, personal fates of affected people.

But "leaving unmentioned" does not mean "ignoring" or "dismissing as negligible." If, for example, I assert an overall positive development for a certain problem on the basis of the relevant statistics, this does not mean that I deny existing negative tendencies. It only means that, according to the available data, the positive tendencies for this issue predominate. To state a positive development also does not mean to tacitly accept the remaining shortcomings – it can rather help to direct the focus of action to these very grievances.

The book is written in this spirit.

Part I

Three Conclusions

2

The Most Important Challenge Is: Switching from Fossil to CO_2-Free Energy Sources

A Dream

Let us dream for a moment at the beginning of this book. We are entering a fictitious world that differs from our real world only in one – but, as we will see, quite decisive – point: In this world, the problem of climate change does not exist. And it doesn't exist because mankind began to seriously curb CO_2 emissions 20 or 30 years ago, so that global warming has been limited to about 1 degree Celsius.[1] In this fictitious world, what are the future prospects up to 2100 (it is hardly possible to look further in a substantiated way) for the planet's ecosystems and for the foundations of human life and economic activity?

To answer this question, I will first briefly summarize the chapters of Parts II-IV, i.e. give an overview of the future projections derived there (cf. Figure 2.1).

[1] Example: If between 2000 and 2010 the large increase in CO_2 emissions (about 10 billion tons) had been avoided by the use of PV, wind and nuclear energy instead of coal in electricity generation, and emissions were reduced to zero between 2010 and 2040, humanity would have emitted as much CO_2 on the whole as it has in reality to date. Such a scenario would have been feasible at relatively limited financial cost.

© Springer-Verlag GmbH Germany, part of Springer Nature 2022

T. Unnerstall, *Factfulness Sustainability*,
https://doi.org/10.1007/978-3-662-65558-0_2

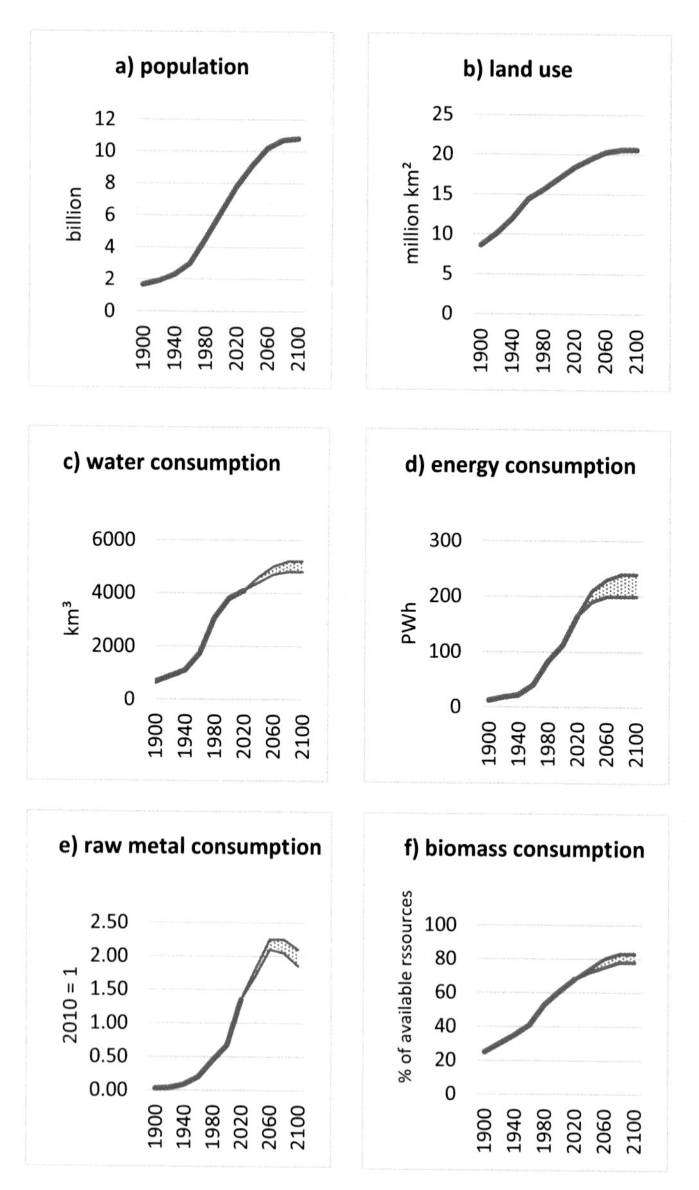

Fig. 2.1 Population (a) and resource consumption – land (b), water (c), energy (d), metals (e), biomass (f) – of the world, 1900–2100; from 2020 forecast values. (metals = iron, copper, aluminum (mean values); sources: see respective chapters, own calculations)

- The **world population** will reach an all-time high of 10–11 billion people in the second half of the century – depending on developments in Africa, either as early as 2060/70 or a few decades later. After that, it will slowly decline again (Chap. 5; Fig. 2.1a).
- The **land area** occupied by humans for intensive use (agriculture, buildings, transport) will also reach a plateau of around 20 million km^2 in this century.

  Humans also use forests for their timber needs, and they let their livestock to graze on a further 30 million km^2, but their direct life and activity will thus continue to be limited to a maximum of 5% of the Earth's surface (Chap. 6; Fig. 2.1b). The earth offers considerable reserves of still undeveloped arable land.
- The **food supply** for 10–11 billion people can in principle be regarded as secure under (largely) stable climatic conditions. Even today, food production is in principle sufficient for the current world population of 8 billion people.[2]

  Moreover, there are no indications that the trend which has now been continuing for 50 years – food production is increasing faster than the world population due to improvements in agricultural techniques – would change in the coming decades. Particularly on the arable land of Africa and South Asia, there is great untapped yield potential (Chap. 7). Soil erosion in many parts of the world is often cited as a threat to food supply of mankind in the future – in my opinion, not rightly so, because on most areas it can be compensated for or contained by fertilizer use and adapted agricultural methods. Moreover, very large areas are still available for additional arable land.

[2] Thus, hunger in the world is essentially a consequence of problems of distribution and inefficient use of existing resources, which are mainly caused by political and social conditions.

- In our fictitious world without (significant) climate change, the current bottlenecks in **drinking water supply** in some parts of the world will remain an essentially local and further decreasing problem. With a few exceptions,[3] they can be resolved with existing technologies – mainly with seawater desalination plants plus transport pipelines in the arid regions of the world and with the construction of basic drinking water treatment in the poorest countries. In particular, there is no global scarcity of resources here: humanity's use of drinking water (including irrigation in agriculture, the main consumer) will increase only moderately in this century, remaining at less than 15% of available renewable resources (Chap. 8, Fig. 2.1c).
- **Energy** resources are abundant on Earth. Even on the basis of already existing technologies – i.e., ignoring future technical innovations – the renewable, CO_2-free energies sun and wind alone can supply tens of times the energy demand of mankind (Chap. 9).

  In addition, current projections indicate that this energy demand will reach a plateau at around mid-century due to progressive energy efficiency. This plateau is likely to be 30–50% above current energy consumption (Fig. 2.1d).
- The amounts of **mineral raw materials** accessible to man (iron, copper, aluminum, phosphorus, lithium, and many others) are almost inexhaustible. But even if one restricts the consideration to the deposits identified today (with relatively high raw material concentrations) and ignores further technical development, the available resources will not become scarce in the twenty-first century (Chap. 10).

[3] The exceptions mainly concern countries whose drinking water supply depends to a considerable extent on transboundary flows, i.e. on developments/decisions in neighboring countries. Political understanding will remain necessary here.

Due to increasing recycling rates, it is highly likely that – despite further global increases in industrial production and construction – most raw material demands will reach a maximum in this century and then slowly decline again: humanity is moving (albeit more slowly than it could) towards a largely circular economy in many areas (Fig. 2.1e).

- The "non-energy" **ecological footprint** of humankind is currently at 70% of the relevant renewable resources. In other words, human agriculture, forestry, and fisheries are sustainable in the sense of this concept.[4] In all likelihood, this will remain so even with 10–11 billion people (Chap. 11, Fig. 2.1f).
- Even though humans actually live and work on a relatively small part of the Earth's surface, the effects of human civilization can be felt right into the far corners of the planet, and this increasingly limits the habitat and living conditions of many plant and animal species. Especially in the second half of the twentieth century, this has led to a massive decline in animal populations and also to an accelerated **extinction of species.**

For some decades now, a trend reversal has been emerging (mainly due to the strong increase in protected areas worldwide and other efforts to protect species).[5] The overall loss of vertebrate biodiversity has slowed down, with the exception of amphibians and, in part, the tropics (Chap. 12).

---

[4] A number of impacts of current agriculture in many countries – species extinction, dead zones in the oceans, groundwater pollution, in some cases soil erosion, inappropriate treatment of livestock – are not captured by the concept of "ecological footprint" and are therefore not included in this statement.

[5] This and the following statements refer exclusively to the well researched and monitored situation with vertebrates (i.e. the more highly developed animals). With regard to insects, no well-founded statements can be made: Neither on the number of species, nor on the natural (i.e. evolutionary) extinction rate, nor on the human-caused extinction of species are there any reliable data on a global scale.

It is difficult to forecast future trends here; but even in a rather pessimistic scenario, no more than 1–2% of today's vertebrate species are likely to become extinct in the twenty-first century – in our fictitious world with a largely stable climate. Although this is an irretrievable loss, it cannot be considered a fundamental impairment for planetary ecosystems or for human livelihoods. Only the coming centuries/millennia will determine whether humans are really causing a new mass extinction in the Earth-historical sense.

- In this fictional world with only moderate climate change, **tropical rainforest deforestation** is arguably the most direct and perhaps the most severe human intervention in nature. It means the immediate destruction of complex ecosystems and the loss of many very specialized plant and insect species. First, however, the problem is limited to relatively small areas of the world overall (less than 0.5% of land area per decade); and second, it has little impact on the living conditions of current and future generations of humanity.[6]

Nevertheless, it is important to note that here, too – after largely unbridled deforestation in the second half of the twentieth century – a trend reversal can be observed (Chap. 13). In recent decades, forest loss rates have declined significantly. It is unlikely that this decline, which has now been going on for many years, will be completely reversed: the opinion of the global public is too determined in this respect, and the pressure on governments and business enterprises involved is already too strong.

[6] It is often argued in this context that the loss of plant and animal species in the rainforest will reduce the huge "gene pool" native to the area, which could be of particular importance for future medical developments. This argument is certainly true, but it is rather theoretical – and it is not about a potential deterioration of future living conditions compared to today, but only about a potential *reduction of improvement* compared to today.

In light of this, we can assume that no more than 10–15% of today's rainforests will disappear by 2050, and that deforestation will end in the second half of the twenty-first century.[7] Then, in 2100, more than 5 million km^2 of untouched tropical rainforest would still be preserved[8] – an area as large as the whole of Europe (excluding Russia).

- **Plastic waste in the oceans** and, as a consequence, plastic consumption is a topic that has often been discussed emotionally in recent times in Europe; and it has been the subject of many political measures as well as intensive social commitment. In fact, however, the impact of untreated plastic waste on both nature and humans is quite limited. An estimated 500 species of marine life are affected (but not threatened in their existence), many coasts are polluted and have to be cleaned up again and again, huge plastic dumps float in the open sea – the images go around the world, but for the most part these are problems that have mostly local causes and local repercussions.[9]

[7] During the last international climate conference in Glasgow in November of 2021, most involved countries (incl. Brazil) have even committed themselves to ending deforestation already by 2030.

[8] This presupposes that with these 10–15% further deforestation – 40–45% of the original rainforest stock would then have disappeared – a "tipping point" is not exceeded after which the ecosystem "rainforest" with its complex cycles of rain-evaporation-cloud formation-rainfall as a whole is endangered. Whether such a tipping point exists or at what point it would be reached has not yet been scientifically clarified.

[9] To be distinguished from this is the problem of "microplastics," which originate partly from the decomposition of plastic waste in seawater, but primarily from other products/processes. There are clearly global impacts and dispersal mechanisms here. Unfortunately, I could not address this issue in the present book because there is too little established knowledge and hardly any global data. In particular, although the relevant research has so far produced no evidence of negative (even longer-term) health effects on humans and most animals, a final result has yet to be reached. This could be an important field of action for environmental policy in the near future.

Moreover, the solution is not far away. The core issue is the introduction of orderly waste treatment in the poorer countries, especially in Asia and Africa. Since all experience shows that this is closely correlated with economic development, significant progress can be expected here in the coming decades (Chap. 14).

Also, the core question regarding plastic waste – incinerate, landfill or recycle? – is certainly worthy of further research and technical development. But in the end, it is of only limited importance. In properly managed landfills, plastic is not a problem waste, and CO_2 emissions from incineration are negligible.

- The environmental problem of **"dead zones in the world's oceans"** is a consequence of the high, often excessive, use of fertilizers in agriculture[10] (Chap. 15). With regard to future prospects, two issues can be distinguished here:

1. The largest dead zones – in the Baltic Sea, North Sea, Black Sea, Gulf of Mexico, East China Sea – have been known for many years, they are the subject of intensive scientific observation, and there are long-term programs to reduce the causative fertilizer inputs. Here it can be assumed that the affected areas will slowly recover over the next decades.
2. In addition, there are about 400 other known, much smaller dead zones; and it is quite likely that there is a high number of unreported cases. It is almost impossible to forecast further developments for this problem area. But even if one assumes that the real number is twice as high and will continue to rise in the coming decades, this only accounts for a total area of

[10] There is a second cause of dead zones: Even with a climate warming of only 1 degree (as is currently the case), naturally existing oxygen-depleted zones in the oceans are expanding. However, the effect is limited to an area well below 0.1% of ocean areas.

well under 1% of the total coastal waters. Moreover, partly because these are generally seasonal rather than permanent dead zones, the ecological consequences are limited: Most animals can escape. The consequences for humans are also manageable; tourism and local fisheries may be affected. In any case, due to its essentially local characteristics, this is actually not a global environmental issue.

- In our fictitious world with very limited climate change, coal-fired power plants are already disappearing and the majority of cars are powered by electric motors. Therefore, **particulate matter** in the air is largely history; and apart from special local circumstances – e.g., desert sand in the Middle East, wood burning in fireplaces – air quality will continue to improve worldwide in the coming decades (Chap. 16).
- Environmental pollution with **heavy metals**, especially mercury, has been declining for decades. This development has been achieved primarily through a series of international agreements and has now led to concentrations in the human body below the limits considered safe, in most regions of the world (Chap. 16). In other words, as far as is currently known, the numerous pollutants that have entered and are entering the environment as a result of human activities do not pose any (large-scale) threat to the living conditions of future generations.

Interim Conclusion

Conducting a thought experiment, we imagine a world in which mankind has already started decades ago to follow up on the scientific findings regarding climate change with actions, and in which therefore the anthropogenically caused global temperature increase is limited to about 1 degree. In this fictitious world, the future prospects for the twenty-first century summarized above, condensed to four statements, look as follows:

1. The planet's natural resources are by far sufficient to ensure the biological basis of life for future generations – food and drinking water – without restrictions.
2. Until 2100 (and far beyond), there are no relevant limits to human activity in terms of energy resources and mineral raw materials.
3. The ecosystems of the planet show clear traces of human activity almost everywhere, and in a number of places the environment is destroyed or polluted for many decades to come. But this does not result in any substantial (i.e. beyond these limited areas) impairment of the living conditions of future generations.
4. The most serious ecological consequences of the almost explosive development of human civilization in the last 100 years are the decline of animal populations and the accelerated extinction of species. But even here, if current trends continue, the tremendous wealth of animal and plant species on our planet will be diminished only slightly in the twenty-first century.

So: What if the dream was true, what if we had limited (or did limit) climate change successfully? There would undoubtedly still be serious ecological challenges. But humanity would have enough time to solve them; many trends are already pointing in the right direction; and there would be no reason to be fundamentally concerned about the living conditions of future generations.

The Reality

So much for the dream. Unfortunately, the real world of 2022 looks different. Humanity's common will to combat climate change has often been invoked, and since 2015 has been cast in the binding Paris Agreement; climate protection is the subject of a multitude of technical developments, political programs, economic activities and social initiatives around the globe. We are going in the right direction, but the speed of change is still not even close to being sufficient. Humanity is still heading for climate change well beyond the 2-degree mark.

In a world with such fundamental climate change,[11] what happens to the four statements mentioned above?

1. The natural resources of the planet – in terms of suitable areas for agriculture, in terms of rain and groundwater – remain largely intact, but the spatial and temporal availabilities change very significantly. Thus, a disproportionately higher technical, financial and also political effort is required to ensure the biological basis of life for all people.
2. *(Unchanged:)* Until 2100 (and far beyond), there are no relevant limits to human economic activity with regard to energy resources and mineral raw materials.
3. Entire regions in the tropics are expected to become practically uninhabitable, i.e. hundreds of millions of people will lose their homes; this has unforeseeable consequences with regard to migration, regional conflicts, security of supply, and much more.

[11] We use a medium climate change scenario here (RCP 6.0). The consequences described are now largely the consensus of climate science and follow-up research.

In many countries of the world, weather extremes – droughts, floods, heat waves, hurricanes – will increase sharply, so that the natural living conditions for large parts of humanity are directly and noticeably deteriorating.
4. For ecosystems, the rapid shift of climate zones means that species extinctions are likely to accelerate significantly compared to the trend of recent decades and reach substantial proportions by 2100.

The conclusion from the summary of this book and from our thought experiment, i.e., from the juxtaposition of the two worlds, is clear. Of all the ecological hotspots of our time, climate change is by far the most serious problem – both for the living conditions of future generations and for the planet's ecosystems.

Ecological Priorities

There are very clear priorities for action with respect to (ecological) sustainability:

First comes the need for action on climate protection. This means above all[12] that the use of fossil fuels must practically be phased out within the next 30–40 years.

In second place, clearly behind, is the successive cessation of (large-scale) rainforest deforestation.

In **third place**, again at some distance, come the issues of further improvement of agriculture (optimization of fertilizer and pesticide use, avoidance of soil erosion); continuation of efforts to protect species (including designation and

[12] CO_2 emissions from fossil fuels (coal, oil, natural gas) account for about 75% of total greenhouse gas emissions worldwide.

better management of protected areas); proper treatment of plastic waste; further reduction of pollutant emissions[13].

Of great, even crucial importance is the time dimension. As a conclusion of Part IV of this book, one can assert (even if it may sound somewhat cynical): One or two decades more or less in rainforest deforestation, in plastic waste discharges into the sea, in fertilizer optimization, in the reduction of pollutants are of course not at all without consequences, but in the end they do not make a qualitative difference. One or two decades more or less in climate protection, on the other hand, could change the entire world.

To put it positively and pointedly: The question is not *whether* the age of fossil fuels will be replaced by the age of renewable and nuclear energies in the course of the twenty-first century, thus reducing CO_2 emissions to near zero. The only question is *how quickly* humanity will achieve this transition.

All the prerequisites are there: Knowledge, technologies, capital, political instruments. We just have to do it.

[13] The extensive avoidance of microplastics should also be mentioned here, despite the lack of knowledge to date regarding negative ecological effects.

3

"The Western Economic System and Way of Life Are Unsustainable" – False!

"Western capitalism and living standards are at the expense of future generations. It is associated with too much consumption of resources, with environmental destruction and with climate change. We must therefore fundamentally change our economic system and our consumption patterns." I am sure that you have come across such statements in one way or another: in books, newspapers, on television or in the social media.[1] In a nutshell, the verdict is: "The Western economic system and way of life are ecologically unsustainable" – and this paradigm has been playing an increasing role in public and political discussions for some years now.

But is the verdict really true?

[1] Examples: "The current, globally defining model of civilization has produced a mass consumption that clearly exceeds the capacities of the earth. The need for profound change is obvious" (Meinert 2018, p. 6). "Sustainable development is only possible with new models of prosperity and consumption patterns" (Schneidewind 2018, p. 54). "With the resources available on Earth, not all people can live at the level of prosperity currently enjoyed by North Americans and Western Europeans" (Steffens and Habekuss 2020, p. 54).

© Springer-Verlag GmbH Germany, part of Springer Nature 2022

T. Unnerstall, *Factfulness Sustainability*,
https://doi.org/10.1007/978-3-662-65558-0_3

The Ecological Sustainability Parameters of the West

In order to answer this question systematically and comprehensively, it is again best to first summarize the core data of the book (Parts II–IV) referring to the West,[2] cf. Figure 3.1.

- The **demographic trends** in the West depend, much more than in other world regions, on future immigration. Without (further) migration from outside, the population in the Western countries would fall sharply in the coming decades and by 2100 would return to the level of the 1970s (Chap. 5, Fig. 3.1a).
- Intensively used **land area** already peaked in the West 50 years ago and has been slowly declining again ever since. In return, forest area is increasing. There is nothing to suggest that this trend could change (Chap. 6, Fig. 3.1b).
- Per capita **food** consumption in the West has been virtually constant for the past 20 years. It is 20% above the world average and is characterized in particular by a high consumption of animal products. However, with one notable exception, only domestic resources are used for this in net terms. In addition, the West also contributes to the food supply of Asia and Africa through its highly productive agriculture (Chap. 7).

  The exception is the EU's soy feed imports from South America, which take up about 5% of the arable land there. They can therefore be linked, with some justification, to some of the (past, not present) rainforest deforestation. These imports have declined significantly in recent years.

[2] Throughout the book, „the West" will, in general, be our abbreviation for the USA and the EU combined.

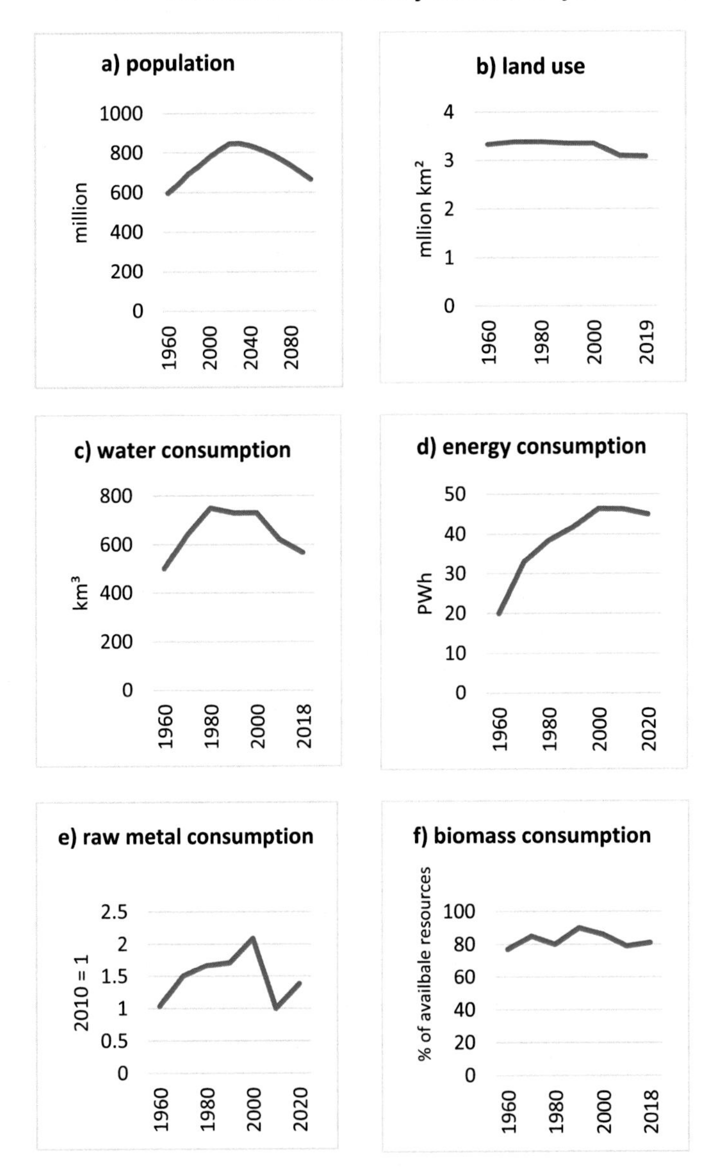

Fig. 3.1 Population (a) and resource consumption – land (b), water (c), energy (d), metals (e), biomass (f) – of the West, 1960 to present. (*without migration from 2020; metals = iron, copper, aluminum (mean values); sources: see respective chapters)

- **Drinking water consumption** in the West has gone down significantly for the past 30 years and is now at less than 15% of the available renewable resources (Chap. 8, Fig. 3.1c).
- **Energy consumption** in the West has also already peaked, and it will continue to fall through 2050[3] (Fig. 3.1d). However, per capita energy consumption in the West is still more than twice as high as in the rest of the world, and the associated CO_2 emissions are correspondingly disproportionately high (Chap. 9).
- Western consumption of **mineral raw materials** has been declining for decades. The main reason for this is increasing recycling rates almost across the entire board; as a result, despite above-average and still rising living standards, per capita raw material consumption of iron, copper, aluminum and phosphorus is now on the same order of magnitude as the world average (Chap. 10, Fig. 3.1e).
- The very large **ecological footprint of** the West – often cited as "direct evidence" of Western economic activity and consumption being unsustainable – turns out, on closer analysis, to be just another way of depicting the high CO_2 emissions of the USA and the EU. Indeed, if one looks at the ecological footprint without the CO_2 emissions, the result is the opposite. The "non-energy" ecological footprint of the West is not too high, it only utilizes about 80% of the available renewable resources, and this with a decreasing tendency (Chap. 11, Fig. 3.1f). In other words, agriculture, forestry and fisheries of the Western countries are sustainable (in the sense of this concept).

[3] The main drivers of this trend are the replacement of two very energy-inefficient technologies - internal combustion engines and fossil power plants – with e-mobility and solar/wind.

- The **extinction of vertebrate species** in the West – already less pronounced than in the rest of the world – has been largely halted in recent decades. Animal populations are now mostly stable, and in many protected areas they are even increasing again[4] (Chap. 12).
- **Forest area** in the West has been increasing steadily (albeit slowly) again for about 100 years (Chap. 13).
- The annual plastic consumption of the West is extremely high, at 150 kg per capita, compared with a world average of 50 kg per capita. However, the responsibility for the environmental problems resulting from plastic use – primarily **plastic waste in the oceans** – lies elsewhere[5]. Plastic waste in the West is subject to modern and systematic waste management; and plastic waste exports (which were never a significant factor) are also largely history (Chap. 14).
- **Dead zones in the ocean**, on the contrary, are primarily a problem for the West in terms of size, according to current knowledge. The US and the EU are responsible for four of the five mainly affected (Baltic Sea, North Sea, Black Sea, Gulf of Mexico). Fertilizer inputs from agriculture – the cause of the problem – have been reduced for decades, but the affected zones are recovering only slowly (Chap. 15).
- **Air quality** in Western countries was the focus of incipient environmental policy 50 or 60 years ago, and much progress has been made since then. The most important remaining problem, particulate matter pollution, will largely disappear as transportation shifts to alternative powertrains (Chap. 16).

[4] The situation for invertebrates, especially insects, may be different. I had to confine myself in this book (Chap. 12,) to the vertebrates (mammals, birds, fish, reptiles, amphibians), because there are no sufficient worldwide data for the other animal classes.

[5] Cf. footnote 9, Chap. 2.

- With respect to other **pollutants** – from PCBs to CFCs to mercury – the West has responded relatively consistently after the respective health hazards became known. In the case of mercury, for example, Western emissions are now close to zero (Chap. 16).

Interim Conclusion

If we put aside the topic of CO_2 emissions and climate change in a first step, we can summarize the above-mentioned points in four statements.

1. Despite continued economic growth and despite further increases in living standards, overall resource consumption is declining in Western countries. Arable land, drinking water, energy, metals, fertilizers, ecological footprint – in all these areas, the consumption figures have been showing falling values for decades.
2. The current generation in the West does not live (in ecological terms) at the expense of future generations: future self-sufficient food production is not significantly threatened[6]; drinking water and biomass use does not exhaust the annually available renewable resources; only tiny fractions of the Earth's existing energy and raw material reserves are being used.

[6] In particular, soil erosion is moderate (also compared to other world regions) and is actively managed in many places (cf. Fig. 7.6)

3. With its current resource consumption, the West does not live at the expense of other regions of the world.[7]
4. Finally, the pressure on ecosystems in the West tends to decrease overall, or at least not to increase further: this can be clearly seen from the indicators of land use, species extinction, animal populations, development of dead zones, pollutant emissions, and others.

A rather clear conclusion. However, two objections are quite obvious:

The **first objection** concerns individual grievances for which the West is (partly) responsible. There are still tropical timber imports from deforestation of rainforests; there is European plastic waste on the beaches of Asia; there are Western companies that are mining raw materials in Africa without adequate environmental standards (and working conditions); there are overfished fishing zones also in the West; a good 10% of all groundwater measuring points in the EU show too high nitrate contamination from agriculture; there are still large "dead zones" in the Baltic Sea; and the list could be extended.[8]

No question: It is good and important to uncover such grievances and to tackle their elimination. Nevertheless, the above four statements remain unaffected by this objection. The states of affairs cited are not *constitutive of* economic

[7] With regard to food and feed, the West is, on the contrary, the world's largest exporter (in total and in net terms). The so-called virtual water imports are, low and structurally uncritical, as they largely originate from countries with abundant water supplies (Chap. 8). The (often cited) plastic waste exports of the West to Asia have always accounted for only a very small part of the problematic plastic waste on coasts and in the sea, but have now largely stopped anyway (Chap. 14). The West's contribution to current global pollutant emissions is small.

[8] One grievance, rightly denounced again and again, concerns the treatment of farm animals, which in many respects is not appropriate to the species. However, this problem is not limited to the West and is more an ethical issue than an ecological one.

activity, consumption and life in Western countries; this can be seen from the fact that these problems largely show a downward trend.

The **second objection** goes deeper and concerns the history. The last 250 years of human history were very much shaped by the culture, economic system and consumption patterns of the West. This epoch has now ended, and the West has largely lost its dominant role. But it is during this historic period that the rapid increase in population, resource consumption, land use, intensity of agriculture, pollutant emissions, and thus all the ecological hotspots of the present discussed in this book emerged. In this sense, the West bears a historical debt, and it therefore now has a particular responsibility to develop and practice ecological sustainability within its sphere of influence. However, it would be wrong to hold it responsible for the current ecological problems in other regions of the world: Their solution is primarily the task of the respective societies and governments.

Thus, while this objection is justified, it does not change the interim conclusion: putting aside the issue of CO_2 emissions, the verdict that "the Western economic system and way of life are ecologically unsustainable" is simply not true in view of the facts and trends of development.

The CO_2 Emissions of the West

But isn't the verdict nevertheless true if we do now take into account the disproportionately high CO_2 emissions of the EU and the U.S., i.e., if one includes climate change in the assessment?

At first glance, the answer "yes" seems inevitable, because three assessments are certain:

- Climate change represents by far the largest and most serious "unsustainability" of human activities; it threatens the living conditions of future generations and many planetary ecosystems alike.
- The West (EU + USA) is not only historically responsible for half of mankind's CO_2 emissions to date, it also has by far the highest current level of per capita emissions (Fig. 3.2).
- The main cause of this, namely the very high energy consumption, is indeed a central feature of the economy and consumption in Western countries; it is constitutive of our way of life: high individual mobility, generous housing, vacations in faraway countries, rich equipment with electrical appliances, energy-intensive industrial production.

At second glance, however, this chain of reasoning has a crucial gap. Energy consumption and CO_2 emissions, i.e. fossil fuels, are *historically* inseparable, but *systematically*

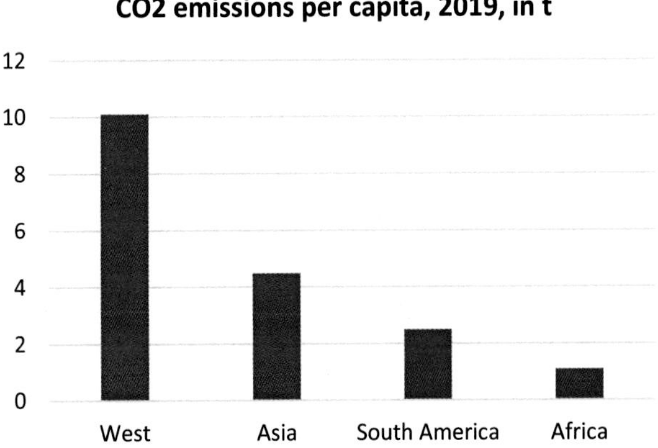

Fig. 3.2 Per capita CO_2 emissions compared; 2018 values. (Source: OurWorldinData)

there are not. If the West generated its electricity from solar, wind and nuclear energies, switched to e-mobility, and ran heating, airplanes, industrial processes on CO_2-neutral PtX energy sources, then its high energy consumption would be quite unproblematic. In particular, such an energy system would be sustainable because only a small fraction of the available renewable energy resources is necessary.[9]

To put it bluntly: the verdict "unsustainable" regarding the western way of life is indeed false.[10] In order to become (ecologically) sustainable, the West does not have to fundamentally change its economic system or its consumption levels, but only its energy system.[11]

The Real Responsibility of the West

Please do not misunderstand me. My point here is not to declare the current free market system[12] as the ultimate wisdom, or to advocate unlimited consumerism. Quite the opposite. Overcoming an understanding of economic success where quantitative growth is the sole indicator, developing a broader concept of "prosperity" and the social objectives in this regard – these tasks, I am convinced, are among the

---

[9] Resources for current nuclear power plants are limited, but they will most likely be replaced by nuclear fusion technology (for which there are de facto unlimited resources) in the next 100–150 years, cf. Chap. 10.

[10] Only the question of environmental sustainability is at issue here. Other (such as social) dimensions of sustainability are not considered.

[11] About 90% of the climate relevant emissions (CO_2, CH_4, NOx incl. LULUCF) of the West result from the combustion of fossil fuels. Of course, in the long run, we have to think about how to avoid or compensate for the remaining 10% of emissions. A number of technical options are already available to this end. The often-discussed reduction of the high meat consumption in the EU and the USA is certainly an issue for many reasons; but even halving it will only bring about a 3% reduction in emissions.

[12] In any case, there are major differences in this system: The US version and the "Swedish model" are worlds apart.

real major challenges facing humanity in the twenty-first century. And I am also certain that important progress will be made on these issues.

My point is simply this: solving these tasks is not a *prerequisite* for the West to become CO_2-neutral, i.e., to pursue ambitious and successful climate protection. Rather, it's the other way around. Mitigating climate change in the sense of the Paris Agreement is a prerequisite for the next generations (not to have to deal with the manifold consequences of drastic climate change, but) to be able to focus their attention on these and other challenges.

Actually, this is good news. Why? Because it follows that, in the end, climate protection is conceptually not so difficult. We don't need a "great transformation" of the economy and society, we don't need lengthy and divisive discussions about consumption patterns, we don't need to argue about air travel or SUVs. All we have to do is replace most fossil fuels with CO2-free energy sources by 2050 – no more, but also no less. We have the necessary technologies to do it, and we can afford it.

This is precisely where the West's main ecological responsibility lies. It will only meet this responsibility if it leads the way into the new energy world and if it helps poorer countries to follow the same path, both technically and financially.

This is also the primary responsibility of our generation to future generations. Yes, today, in the West, we do live at the expense of our children and our children's children. Not because we believe in free markets or because we consume too much, but because we are emitting far too much CO_2 against our better judgment – because we have so far been too lazy to change our energy supply at the required pace.

We do not have to leave an ecologically perfect world to future generations. But we do owe them natural living conditions in which they can freely develop their economic system, their way of life and their culture as they see fit.

4

The Fight Was Worth It

1970: Foundation of the Environmental Protection Agency, USA.
1970: Foundation of the world's first Ministry of the Environment, Germany.
1971: Foundation of Greenpeace, Canada.
1972: Publication of "The Limits to Growth," USA.
1972: First UN Conference on the Environment in Stockholm; foundation of UNEP.
1973: Washington Convention on International Trade of Endangered Species.

The first half of the 1970s can probably be called the "birth epoch" of environmental protection, both in the sense of government action and in the sense of civil society involvement. Some initiatives go back even further – for example, the founding of the World Wildlife Fund in Switzerland in 1961 or the Clean Air Act in the U.S. in 1963 – but environmental protection did not become a broad movement and a firmly anchored pillar in politics until the years 1970–1975.[1]

[1] In September 1970, in a survey of the German population, only 41% knew the term "environmental protection"; a little over a year later, 92% did.

© Springer-Verlag GmbH Germany, part of Springer Nature 2022

T. Unnerstall, *Factfulness Sustainability*,
https://doi.org/10.1007/978-3-662-65558-0_4

In the first 10–15 years, the environmental movement and policy were largely limited to Western countries, but in the course of the 1980s, ecological problems increasingly became the subject of worldwide conferences and agreements: 1986 first whaling moratorium, 1987 Montreal Convention to protect the ozone layer, 1989 Basel Convention to regulate waste exports, 1992 UN Conference on the Environment in Rio de Janeiro, 1992 first Framework Convention on Climate Change.

The spectrum of issues also broadened rapidly. In the 1970s, the focus was on three main challenges:

- Pollution of air and water,
- Threat to animal species,
- risks associated with the use of nuclear energy.

At the Rio-conference in 1992, with 172 nations participating, almost all of the issues discussed in this book can already be found on the agenda. In Rio, the overarching goal of "sustainable development" also became the focus of discussion for the first time.

The development of environmental awareness and the resulting action, which is only very briefly alluded to here, did not happen out of nowhere. It is closely linked, on the one hand, to the then very acute pollution of air and water in the industrialized countries (cf. Chap. 16), on the other hand to the drastic aggravation of practically all global ecological problems in the second half of the twentieth century.

If you look at the corresponding figures in this book together, you will see this immediately: Water consumption (Fig. 8.1), energy consumption (Fig. 9.2), fish catch (Fig. 12.9), species extinction (Fig. 12.1), rainforest deforestation (Fig. 13.2), fertilizer consumption or dead zones (Fig. 15.1 resp. 15.3), mercury emissions (Fig. 16.3) – all these indicators show a steep upward curve from 1960 onwards and take on hitherto unimaginable values.[2] In 1985, therefore, in view of the data available at that time, the fear was not unfounded that these developments could accelerate further and at some point lead to devastating consequences (as also predicted in "The Limits to Growth").

Nevertheless, the road for the environmental movement has often been arduous. To be sure, measures that could be justified in terms of preventing *immediate health hazards* to people usually found acceptance relatively quickly[3]: measures to clean the air, to improve water quality, to fight the ozone hole. But all those measures which were committed to *longer-term/indirect* ecological issues – resource depletion, rainforest deforestation, biodiversity, climate change, dead zones – usually faced strong opposition. In most cases, it was tangible economic interests that opposed environmental concerns; and this clash between ecology and economy continues to shape many environmental discussions today.

The struggle for a balance between environmental protection and economic development, for an optimal coexistence of man and nature, had many protagonists and forms: Demonstrations, citizens' initiatives, green parties, NGOs, environmental politicians in established parties, environmental departments in large corporations, representatives of indigenous peoples. It has been long, it has been tough, and progress has often seemed slower than the growth of the problems.

But the data and figures in this book show that the struggle was worthwhile. If we take the main environmentally relevant indicators as a basis and compare the values from

---

[2] Similar curves could be drawn for the ozone hole, SO_2 emissions, and many others.

[3] The measures to fight the Corona pandemic are another particularly drastic example of this.

1960 to 1990 with those of the first two decades of the twenty-first century, a clear picture emerges (Fig. 4.1)[4]:

- The global rate of natural resource depletion in biomass and drinking water has slowed significantly (Fig. 4.1b, c);
- Species extinctions and declines in animal populations are at significantly lower levels (Fig. 4.1e, f);
- deforestation of rainforests has halved (Fig. 4.1d);
- global mercury emissions have declined (Fig. 4.1g); and many pollutants were already completely banned in 2000.

While none of these global environmental challenges can be considered solved, the outlook for the future is much more positive today than it was in the 1980s – just as it is presented in this book.

To put it another way: The civilizational development of humankind since 1960 has led not only to significantly higher standards of living and levels of consumption in large parts of the world, but also, in parallel, to an environmental awareness and a level of environmental policy that the pioneers of the environmental movement 50, 60 years ago could only dream of.

Of course, this is still not enough. Environmental protection and, even more so, climate protection still do not have the status they need in the global perspective. The commitment of citizens, volunteers, NGO employees, environmental politicians, sustainability-oriented companies and start-ups, and "Fridays for Future" activists must go on. The battles between commercial interests/development needs and environmental concerns must be fought again

[4] One exception is the problem of untreated plastic waste, some of which ends up in the oceans. Since plastic consumption in Asia and Africa did not pick up speed until after 1990, this is a new problem for the twenty-first century. A comparison over time regarding the dead zones is not possible, since no sufficient data are available.

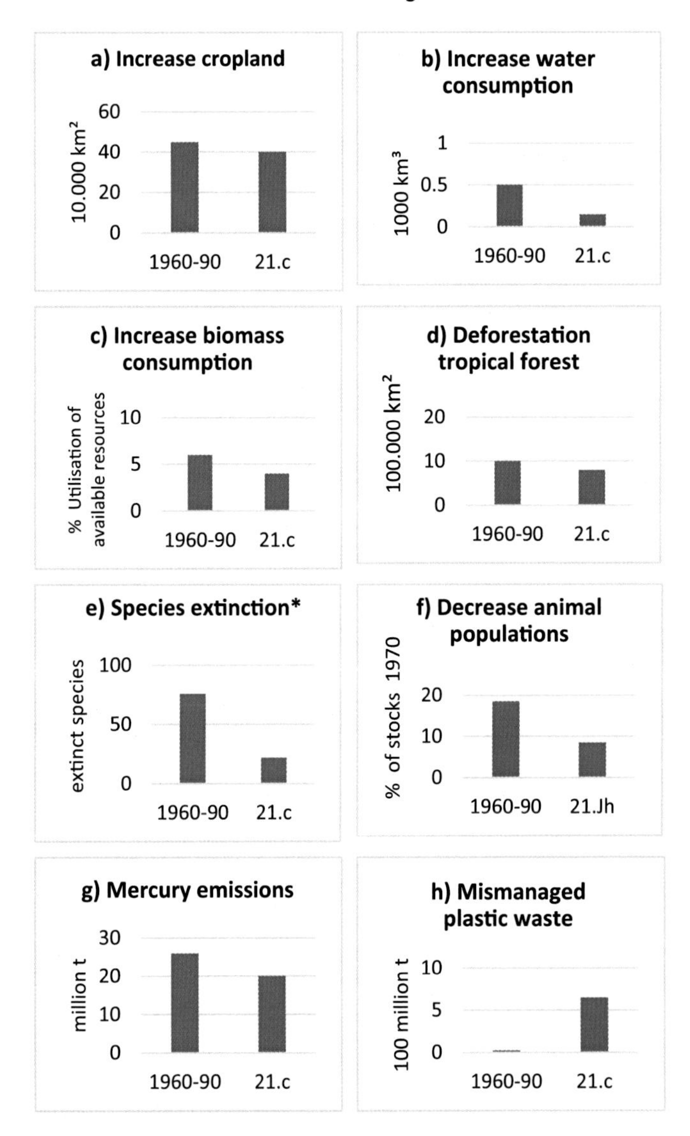

Fig. 4.1 Global ecological indicators, 1960–1990 compared to 2000–2020; values per decade. (*referring to vertebrates; sources: see respective chapters)

and again, and the struggle to reconcile ecology and economy still has a good way to go.

But if there is one conclusion regarding the ecological challenges facing humanity that I myself have drawn from the work on this book, it is this: Pessimism about the future or even doom and gloom – as understandable as they may sometimes be in the face of reoccuring environmental catastrophes and ecological sins – are not objectively justified. Yes, there is a whole range of primarily local/regional problems, and there is the major global threat of man-made climate change. But there are also solutions to them, and humanity has already made significant progress in a whole range of fields.

More generally speaking: Humanity makes mistakes, sometimes big mistakes, but it is also capable of learning and taking appropriate action. And if someone despairs of the slowness in learning or the hesitancy in translating knowledge into action, then he may consider the age of mankind. If the biologists are right, the average life span of a mammal species on earth is about 1 million years. Seen in this light, the species Homo sapiens, the modern human being, with its approximately 70,000 years – projected onto a human life span – is just 6 years old. Indeed, mankind is still young, still in its early childhood, so to speak, still at the beginning of its journey. It has to work earnestly on the tasks ahead of it and on itself – but it can, *we can* go into the future without fear.

You don't believe it? Then read the following chapters.

Part II

Basics

5

World Population

Introduction

Concerns about disastrous world population growth are over 200 years old. In 1798, the famous British economist Robert Malthus published his "Essay on the Principle of Population." Malthus' chain of reasoning in this book was seemingly compelling:

- With a population of about ten million people, a British couple in Malthus' time had an average of 4–5 children, i.e., the so-called "fertility rate" of a British woman was over 4 at the end of the eighteenth century.
- As a result of the onset of the Industrial Revolution in Great Britain, there was a) a marked improvement in living conditions for large parts of the population; and b) associated with this, an increasing decline of the factors that had previously kept population growth in check: infant mortality, violence, disease.
- A fertility rate above 4 and higher life expectancy mean: The population doubles in each generation. Thus, according to Malthus, it would grow exponentially in the future.

© Springer-Verlag GmbH Germany, part of Springer Nature 2022
T. Unnerstall, *Factfulness Sustainability*,
https://doi.org/10.1007/978-3-662-65558-0_5

- However, food production can only grow linearly, never exponentially with a factor of two like the population.
- The inescapable consequence, Malthus concluded, would be hunger and poverty again for most people in Great Britain – and, with the same basic logic working, for the world population as well.

This consequence is still referred to today as the "Malthusian catastrophe."[1]

Indeed, if the fertility rate in Great Britain was still at 4 today, there would be several billion people living on this relatively small island – which, of course, would be impossible. De facto, however, the fertility rate in Great Britain is 1.9, the population is about 67 million, and the country can easily feed itself.

What happened? Why did the Malthusian catastrophe fail to materialize?

Malthus' mistake[2] was not so much that he could not foresee the invention of artificial fertilizer some 100 years later.[3] Rather, the main error was that he failed to see or misjudged the connection between standard of living and

[1] One hundred seventy years later, the renowned American biologist Paul Ehrlich published the international bestseller "The Population Bomb," in which he made very similar predictions.

[2] If we speak here of Malthus's mistakes, this is not to disparage his scientific merits in any way. On the contrary: To have been the first to systematically analyze the relationships between population development, food production and standard of living is an intellectual achievement that can only command respect. Incidentally, it is one of an impressive list of cultural breakthroughs in the second half of the eighteenth century, many of which we still draw on today. In politics (the French Revolution of 1789, the American Constitution of 1776), in economics (besides Malthus, especially Adam Smith's "The Wealth of Nations" in 1776), in philosophy ("Critique of Pure Reason" by Immanuel Kant in 1781), in technology (invention of the modern steam engine by James Watt in 1769) or, of course, in art (major works by Mozart, Goethe, Schiller, Voltaire and many others).

[3] Mainly due to the use of artificial fertilizers, grain yields in today's Great Britain have increased by a factor of about five compared to the time of Malthus: from 1-2 t/ha to 6-8 t/ha.

birth rates: Higher living standards do not lead to rising fertility rates, but in reality to falling fertility rates.

This insight is at the same time the key to understanding the development of world population – today and in the foreseeable future.

The Facts/Projection Until 2100 – Global View

The decisive parameter for population development in a country/the world is indeed the fertility rate (FR) – i.e. the average number of children per adult woman: if it is permanently around 2.1[4] the population is constant. If it is permanently below 2.1, the population decreases; if it is above 2.1, the population increases.[5]

If we then depict the development of this crucial parameter FR in important world regions over the past 60 years, we obtain Fig. 5.1.

This illustration shows that

- the populations in North America and Europe will decline (apart from migration), as the FR has been permanently below 2.1 for more than 40 years;
- the FR in South America reached the 2.1 mark a few years ago and the population there will therefore peak in about one generation (around 2050);

[4] The value is slightly higher than the immediately obvious value of 2.0, in order to reflect the effects of infant mortality, sex ratio (girls and boys are not born exactly in the ratio of 50:50, but in the ratio of 49:51) and others. It also follows that the value is actually country-specific, since infant mortality varies from country to country: For example, the exact value for the EU is slightly below 2.1, while in Africa it is slightly above 2.1.

[5] Of course, strictly speaking, for individual countries this is only true if there is no significant (net) migration in or out.

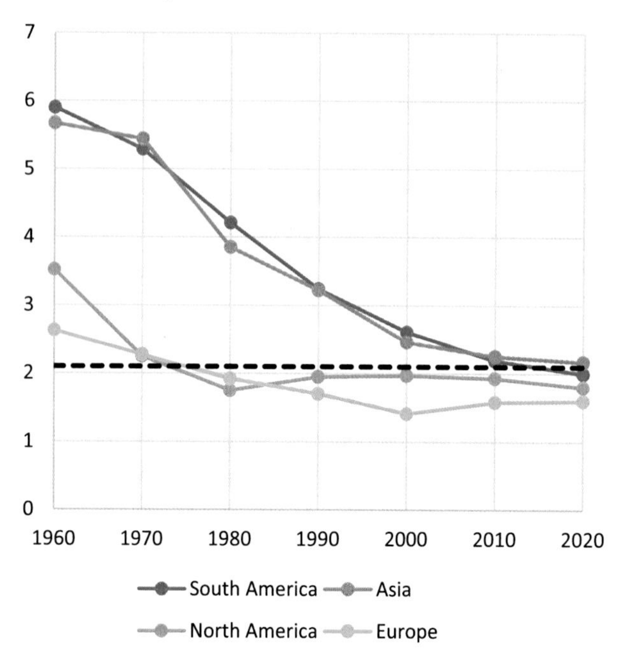

Fig. 5.1 Development of fertility rates on the continents, 1960–2020, excluding Africa and Australia. (Sources: Worldometer, UN World Population Prospects)

- the FR in Asia, with a falling trend, is currently at 2.2, and according to current projections will reach the 2.1 mark between 2025 and 2030, and thus the Asian population will peak about one generation later, around 2060.

On this basis, the development of the world population – with the exception of Africa – between 1960 and 2100 will very likely be as shown in Fig. 5.2.

It is quite certain, then, that the world population excluding Africa will be roughly back to its current level in 2100: in the order of 6–7 billion people.

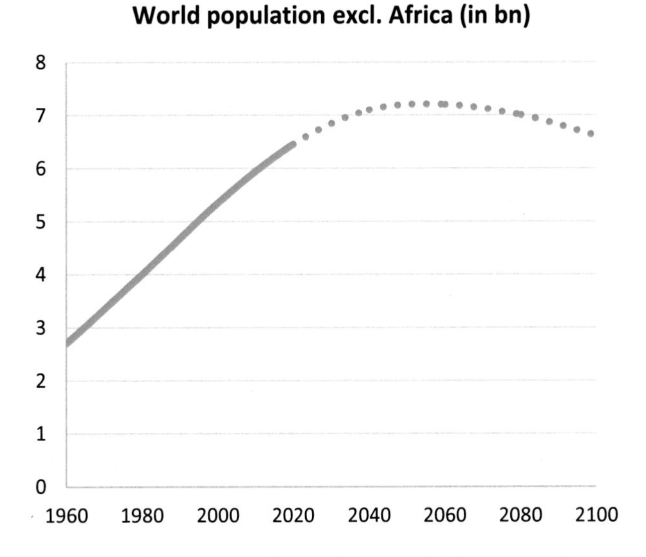

Fig. 5.2 Development of world population outside Africa 1960–2100; from 2020 forecast. (Source: UN World Population Prospects, middle scenario)

While these UN projections can be considered quite certain, population trends in Africa are much harder to predict: Africa is the big uncertainty factor regarding world population in this century. It is true that, also in Africa, FR has declined significantly over the last four decades (Fig. 5.3).

But at the moment, the African population is still growing very strongly, and therefore the current forecast for the future is as shown in Table 5.1.

However, this forecast is only reasonably reliable up to about 2050. For the time after that, the crucial question is how fast the economy and thus prosperity in Africa will grow over the next 30 years.

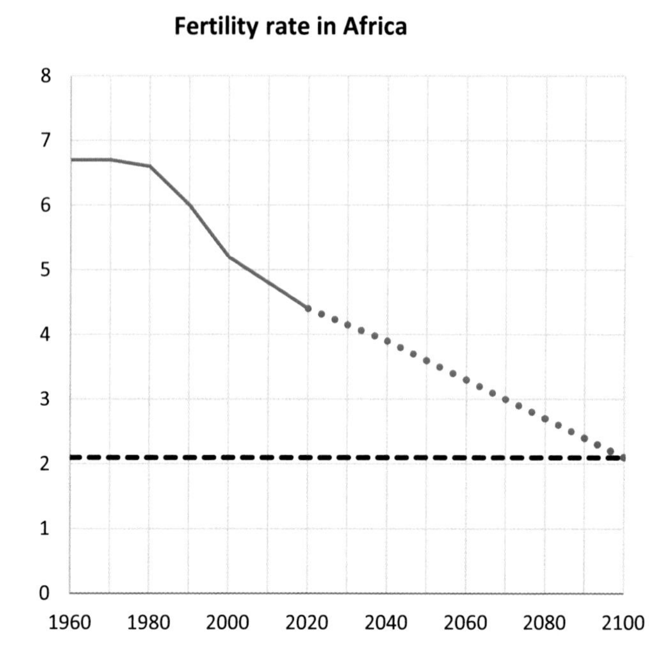

Fig. 5.3 Development of fertility rate in Africa, 1960–2100, from 2020 forecast. (Source: UN World Population Prospects)

Table 5.1 Population development in Africa (from 2050 forecast) in millions

1960	2000	2020	2050	2100
300	800	1350	2500	4300

Source: UN World Population Prospects

As indicated at the outset, material living standards are the key determinant of FR. This can be seen very impressively in Fig. 5.4, which shows the relationship between economic output (GDP) per capita and FR for the 40 most populous countries in the world.

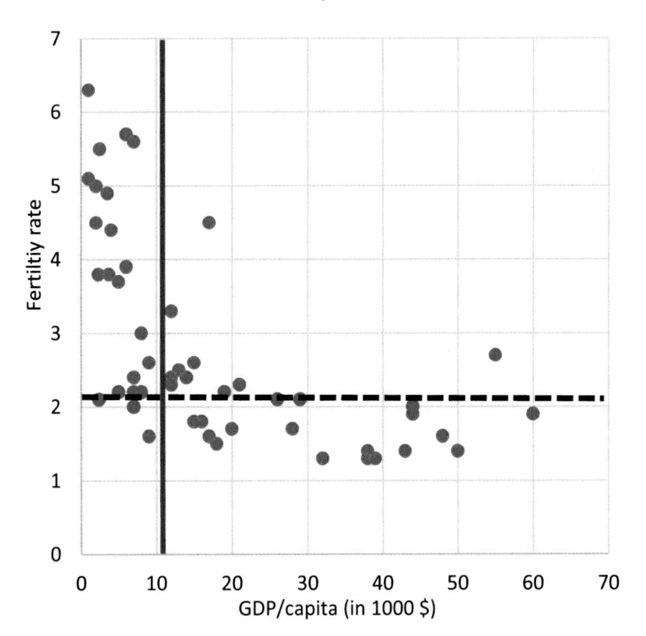

Fig. 5.4 Relationship between fertility rate (FR) and gross domestic product (GDP) per capita (in US$1000 according to purchasing power parities); shown are the 40 most populous countries in the world; 2018 data. The two "outliers" are Iraq on the left and Saudi Arabia on the right. (Sources: World Bank for GDP, Worldometer for FR)

Above a GDP per capita of about US$12,000 per year[6] (by purchasing power), the FR is close to or below 2, i.e., the population is no longer growing. At lower levels of prosperity, on the other hand, it is almost consistently above 2. It is quite striking that this regularity seems to apply

[6] In most countries of the world, household incomes account for about 50–60% of GDP. Therefore, a GDP of US$1000 per month corresponds to an average household income of about US$ 500 per month/capita (by purchasing power).

completely independently of tradition, prevailing religion, form of government and social structures.

There is a second global trend that is important for the future of humanity and the planet: urbanization. The further growth of the world's population will take place exclusively in the cities, not on the countryside. Currently, out of 7.8 billion people, about 3.4 billion live in rural areas and 4.4 billion in cities. For the rural population, this means that the peak has already been reached: From 2020, it will start to decline again. By 2050, it is highly likely that only around 3 billion people will still live in rural areas and just under 7 billion in cities. This development is taking place in a very similar way in all regions of the world – cities will continue to grow all over the globe. This poses new challenges for public transportation, energy and water infrastructure, public safety, and much more.

The Facts/Projection Until 2100 – the West

The population development in the West (EU + USA) in the next decades is difficult to forecast, because it strongly depends on the future extent of migration. Without further immigration (and from today's perspective), it would look as shown in Fig. 5.5; population numbers would then fall back to the level of the 1970s by 2100. However, current projections assume that the population decline caused by low FR will be mitigated (EU) or even significantly overcompensated (USA) by substantial immigration.

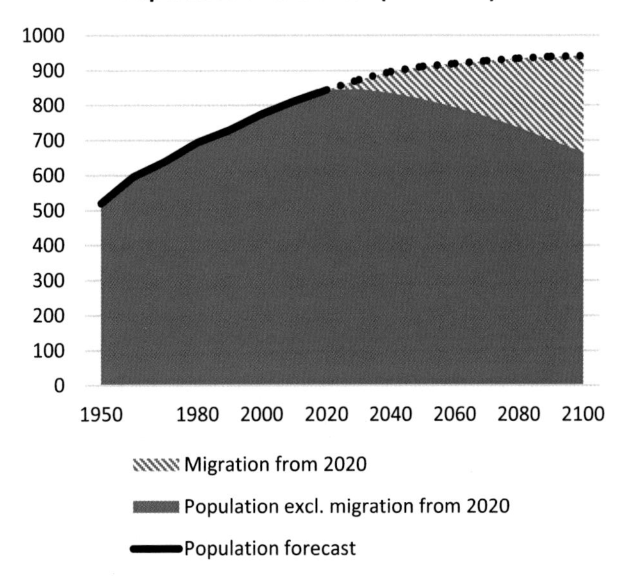

Fig. 5.5 Population development of EU + USA 1950–2100; from 2020 projection. (Sources: UN World Population Prospects, US Census Bureau, ESTAT, own calculations)

Evaluation/Summary

Strong economic growth and the associated significant rise in living standards in all parts of the world except Africa over the past 60 years have led to a sharp decline in birth rates worldwide. The average number of children per woman is now close to or below 2 everywhere except in Africa, which is equivalent to stable or slightly declining population numbers in the long term.

The consequence is simple, but of considerable historical significance: In the second half of this century, the world population will peak at 10–11 billion people – either around 2100 or a few decades earlier, depending on developments in Africa. After that, it will slowly decline again.

For all further considerations in this book, I will therefore use this figure of a maximum of 10–11 billion people as a basis, especially when it comes to the future demand for food, energy and raw materials. In Chap. 7 we will see that the earth's resources are easily sufficient to feed these 10–11 billion people. The "Malthusian catastrophe," i.e., an unmanageable population explosion, will not occur on a global scale.

Excursion: Population Development in Africa

At present, Africa has a population of about 1.3 billion, and it is virtually certain that (apart from possible major migration movements) it will be over 2 billion in 2050.

This is not a problem to begin with: with 2 billion people, the population density in Africa (leaving out the Sahara as uninhabitable) is only about 100 people per km^2. This is still far below the values in Asia (150) or in Western Europe (180).

The land potential available for agricultural use (at least 0.2 ha/capita[7]) will be sufficient to feed this population using modern farming methods: In Western Europe, there is only 0.18 ha/capita, in Asia even only 0.13 ha/capita. In other words: Africa is big enough to be home to 2 or even 3 billion people.

But for the economic and social development of the continent, it would probably be better (if it is permissible to make a judgment here at all) if the fertility rate of 4.4 today were to continue to decline rapidly and approach the

---

[7] See Chap. 6, according to which there is much unused potential arable land in Africa.

stability threshold of 2.1. How long will this take? As recently predicted by the UN, another 80 years? The example of Iran shows that it could also go much faster: In Iran, the FR fell from 4.5 to 2 in just 10 years (1990–2000).

Of course, Africa is an extremely heterogeneous continent – there are worlds between countries like Morocco, Burkina Faso and South Africa. However, for many countries, especially in sub-Saharan Africa, it seems that, in addition to the longer-term development of the economy and standard of living (see main text), a second factor plays a role in the short term for the future development of birth rates: education regarding and the availability of contraceptives. If rapid progress can be made here (also with outside help), this could contribute significantly to a more moderate growth of the African population.

6

Land Use

Introduction

Will you participate in a little experiment? What do you estimate is the percentage of the earth's surface (including the oceans) that has been fundamentally transformed by humans to date, i.e. used for buildings, traffic routes, cropland and mining? Is this

- 5%
- 10%
- 20%?

The correct answer will probably surprise you: It is less than 5%.

Indeed, our intuition often deceives us when it comes to the dimensions of our planet. The earth is large, it has a surface area of 510 million km^2; that is 53 times the area of the USA. Most of it cannot be (permanently) inhabited or cultivated with current technologies (Fig. 6.1):

- 360 million km^2 = oceans
- 30 million km^2 = deserts, mountains, areas with sparse vegetation

© Springer-Verlag GmbH Germany, part of Springer Nature 2022

T. Unnerstall, *Factfulness Sustainability*,
https://doi.org/10.1007/978-3-662-65558-0_6

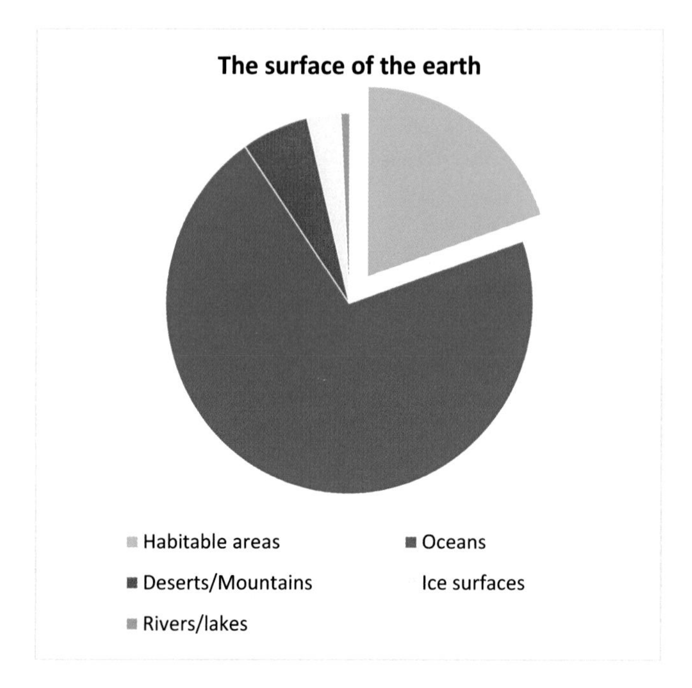

Fig. 6.1 Breakdown of the Earth's surface into basic categories, 2017. (Source: OurWorldinData (Land Use))

- 15 million km^2 = ice areas (especially Antarctica) and glaciers
- 4 million km^2 = rivers and lakes

This leaves only 100 million km^2 – about 20% of the earth's surface – that can be used more intensively by humans. How is this land actually used?

The Facts – Global View

About half of the land area that is in principle habitable is forest (40 million km^2) and areas with mixed vegetation (10 million km^2). Another third (33 million km^2) are also landscapes with mostly low, loose vegetation used as pasture for human agriculture (Fig. 6.2).

Only the remaining ca. 20 million km^2 are actually intensively inhabited or cultivated by humans:

- 16 million km^2 = cropland (incl. orchard plantations, etc.)
- 2 million km^2 = cities, traffic infrastructure.
- <1 million km^2 = raw material mines[1]

Essentially, these uses of land by humans have occurred at the expense of forest: Of the original[2] 60 million km^2 of forest on Earth, ca. 20 million km^2 (cf. Chap. 13) have been

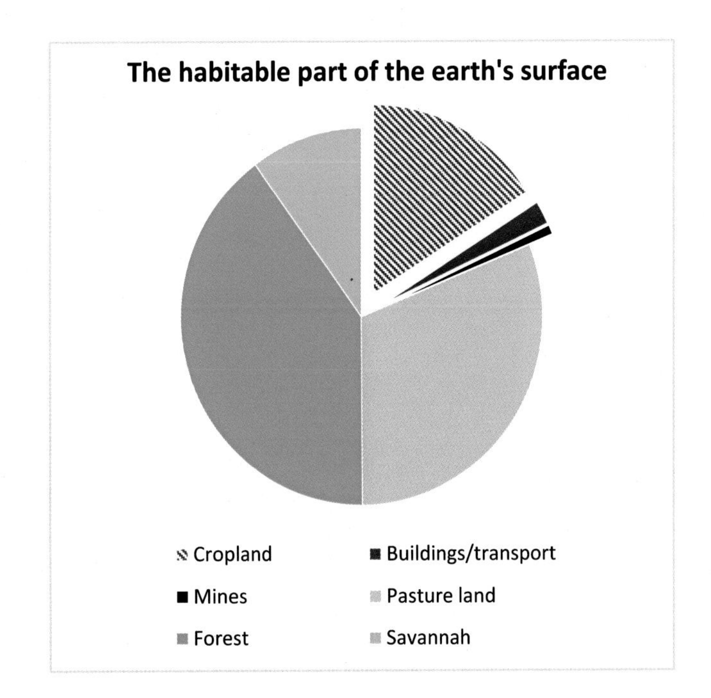

Fig. 6.2 Breakdown of habitable land into land uses, 2017. (Sources: FAOSTAT (Land Use), EU 2018, World Atlas of Desertification)

cleared – starting as early as 3000 years ago, but especially in the last 500 years – to make room for agriculture and settlements.

If we take a closer look at the historical development of human land use over the last 400 years (Fig. 6.3), we can see that new areas were "put under the plow" especially at the beginning and in the middle of the last century; since then, the development of new cropland has declined sharply.[3]

It is important to note that humanity is far from exhausting the potential of arable land that the earth offers[4]:

Total potential cropland =	44–46 million km²
of which forest and protected areas =	18 million km²
of which already used as cropland[5] =	13 million km²
= > potential for further cropland: =	13–15 million km²

Even if protected areas and forest areas are left aside, there are currently well over ten million km² that mankind could still use for crop farming.[6] Larger undeveloped areas exist mainly in Africa (>4 million km²), South America (>3 million km²), China (>1 million km²), North America and Russia.

[1] This area figure is based not only on the actual mine areas; buffer zones with a radius of 5 km around the mine in each case are also included.

[2] The extent of forest cover on the earth's solid surface has been subject to strong fluctuations in the course of the earth's history. By "original" we mean here: after the end of the last ice age about 12,000 years ago.

[3] Exactly the same development can be observed for pasture land: Here, too, new areas were occupied by livestock mainly between 1900 and 1960; for the last 20 years, the area under pasture has actually been declining slightly.

[4] FAO (2011); Zabel et al. (2014). By total potential cropland, we mean here the land with soil qualities "prime" (very suitable) and "good" (well suited) as reported in the 2011 FAO baseline study.

[5] Crop farming is therefore also practiced on approx. three million km² of land that is actually less suitable.

[6] Some of this would be at the expense of current pasture, some at the expense of other areas with mixed vegetation.

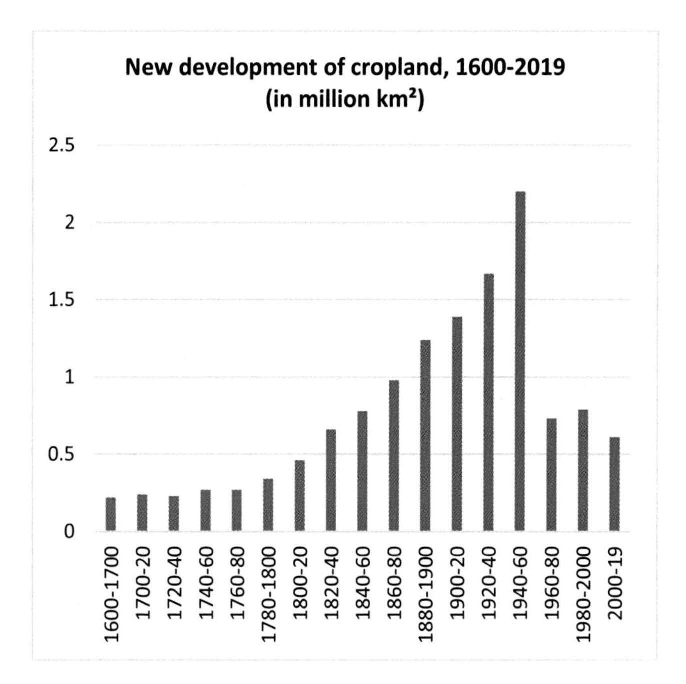

Fig. 6.3 Temporal development of newly tapped cropland 1600–2019. (Sources: FAOSTAT (Land Use), OurWorldinData (Land Use), own calculations)

Climate change[7] will not change this potential significantly by the end of the twenty-first century, according to the current findings of climate scientists.[8] The areas suitable for crop farming will even increase slightly; on the other hand, the overall quality of the soils will decline somewhat.

[7] If I use the expression "due to climate change" here and generally in this book without further specifications, the question arises what extent of climate change I am referring to: after all, it is not at all clear today how fast mankind will reduce greenhouse gas emissions and to what values the expected (further) temperature increase can be limited as a consequence. I base this book on a medium scenario (RCP 4.5 or RCP 6.0 in scientific terms) with a temperature rise of 2.5–3 degrees by 2100: a scenario, in other words, in which the goals of the Paris Climate Agreement are not achieved, but considerable efforts are nevertheless made to combat climate change.

[8] Zabel et al. (2014).

From a regional perspective, the picture is naturally quite different: Especially in Canada, Russia and China, areas suitable for crop farming will increase compared to 2020, while Africa will lose potential in the course of climate change. This will almost certainly exacerbate existing regional problems (cf. Chap. 7), but Africa as a whole could still easily double its cropland despite climate change, without clearing forests or encroaching on protected areas. The same is true for large regions of North and South America.

The Facts – The West

In the West, the peak of intensive land use already occurred 50 years ago: since 1970, arable land has continuously decreased by about 15% (Fig. 6.4). In return, land use by cities and traffic routes has increased moderately, and forests have regrown.

Projection Until 2050

The available studies[9] agree that, from now on, cropland will increase only moderately: About one million km^2 could still be added.[10] After that, no further significant increase is

[9] FAO (2012), Molotoks et al. (2018).

[10] This is a net consideration: due to soil erosion, it is expected that up to two million km^2 of current arable land will be lost to crop farming in the next 40–50 years, i.e., will have to be put to other uses (cf. Chap. 7, Excursion). It can therefore be expected that a total of up to three million km^2 of arable land will be newly developed.

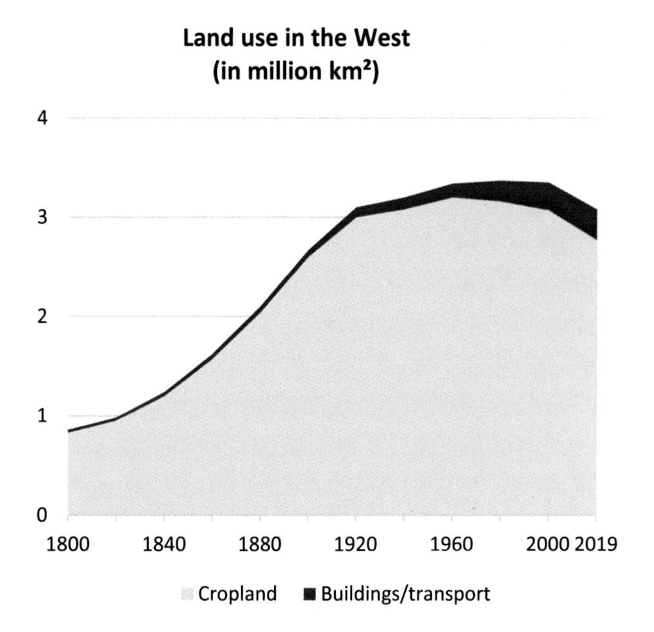

Fig. 6.4 Development of intensively used land in USA + EU, 1800–2019. (Sources: OurWorldinData (Land Use), Globalfootprint (Build up Land), own calculations)

to be expected; the FAO even assumes slightly declining values for the second half of the twenty-first century (cf. Fig. 6.5).

If we also take into account that cities will grow to about twice their current size by mid-century (cf. Chap. 5), the conclusion is clear.[11] At least in this century, the intensively

[11] The area used for raw material extraction will not increase significantly with the foreseeable end of the coal age.

[12] This applies except for possible uses for PV and wind plants, cf. Chap. 9

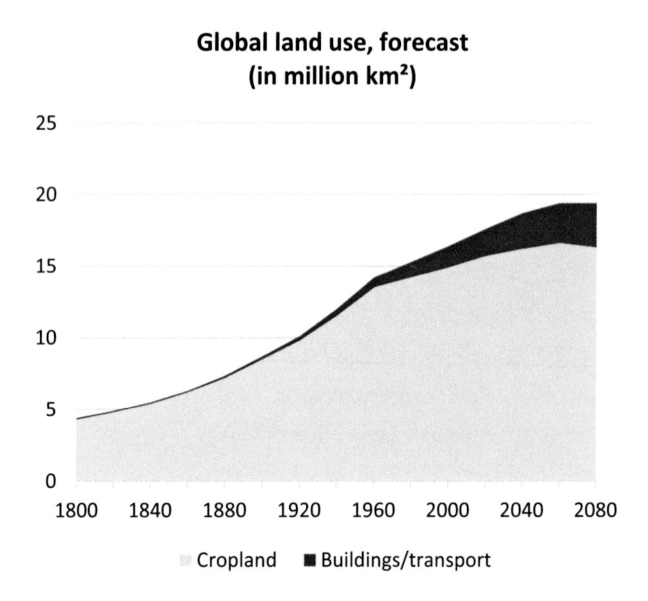

Fig. 6.5 Evolution of the world's intensively used land 1800–2080, from 2020 projection. (Sources: OurWorldinData (Land Use), FAO 2012, own calculations)

used land area transformed by humanity in the above sense will not exceed 20–25% of the habitable regions of the earth and thus 4–5% of the earth's surface.[12]

Evaluation/Summary

In recent centuries, mankind has explored the far corners of our planet, and today vacation offers and adventure trips take us to the most remote regions of the world.

In their daily life and work, however, today's 8 billion people are spread over a fairly small area: their immediate sphere of influence is limited to less than 5% of the earth's surface. And their farm animals can be found on another 6%.

From today's perspective, it is very unlikely that this basic picture will change in this century, despite the projected growth of the world's population. Cities will continue to expand, and in some countries forests will still be cleared (actually unnecessarily so) in favor of farmland, but on a global scale these are developments in the range of less than 1% of the earth's surface.

Thus, contrary to the pictures and opinions often conveyed by the media, humanity is not even close to exhausting the planet's resources in terms of arable farming and food production. According to current knowledge, climate change will not fundamentally change this overall conclusion: while likely to bring about (moderat) regional shifts of land resources, it will hardly alter their global extent.

7
Food

Introduction

The year 2015 arguably marks the high point so far in the international collaboration with respect to the key challenges facing humanity. First, the Paris Climate Agreement was signed by virtually all nations in December 2015, and second, a UN summit was held in September 2015 at which the "2030 Agenda for Sustainable Development" was unanimously adopted.

At the core of this agenda are 17 so-called "Sustainable Development Goals" (SDGs), which all countries of the world have agreed upon. Goals 1 and 2 are: By 2030, extreme poverty and hunger in the world should be overcome. Very deliberately, these two goals come first. Adequate, healthy nutrition is a prerequisite for all forms of human development and activity – for a life in dignity.

Malnutrition has been a constant threat to people in all parts of the world for thousands of years. Even in Europe, as recently as 100 years ago, there were massive problems in supplying the population with food in a number of regions; and in its first ever estimate of the global situation, the UN concluded in 1947 that over 40% of the then approx. 2.5 billion people did not always have enough to eat.

© Springer-Verlag GmbH Germany, part of Springer Nature 2022

T. Unnerstall, *Factfulness Sustainability*,
https://doi.org/10.1007/978-3-662-65558-0_7

Since then – i.e. in the last 75 years –, on the one hand the situation has improved dramatically. Despite the increase in the world population to almost 8 billion people, the proportion of undernourished people today is estimated at only about 10%. On the other hand, we are still talking about more than 800 million people, equal to the population of the EU and the USA combined. This is indeed terrible and explains the high priority given to the issue in the UN 2030 Agenda.

Against this background, is the goal of producing enough food for all 8.5 billion people by 2030 realistic? And can the earth also feed 10 or 11 billion people well and securely later in the century?

The Facts – Global View

For adequate nutrition, humans need on average – across all climatic zones, age groups and activities – about 2400 kcal/day. Since losses in the order of 10% can hardly be avoided, 2600 kcal/day or just under 1 Gcal[1] per person per year must be provided. The current world population (7.9 billion) thus needs about 7.5 billion Gcal of food. Accordingly, if the SDG-agenda goal is to be achieved, about 8 billion Gcal of food must be available in 2030.

If we look at the global production data of recent years with these figures in mind, the emerging picture might surprise you. Currently, 12–13 billion Gcal of arable products are harvested worldwide, broken down as follows[2]:

[1] 1 Gcal = 1 million kcal.

[2] All data in this chapter are taken from the – very comprehensive and well-structured – FAO database (FAOSTAT).

Share of total (%)	Gkcal (in bn)	Category of use
50	6.5	Vegetal foods
25	3.0	Feed; from this, together with green fodder (pastures) 1.5 bn Gcal animal foods are being produced
20	2.5	Other uses: energy crops, índustrial oils, seed
5	0.6	Losses

Thus, the current supply of the world population with food is in the order of 8 billion Gcal. In other words, more than enough food is produced globally; statistically, each person has about 2960 kcal available per day. Table 7.1 shows the more detailed breakdown.

A look at history also shows (Fig. 7.1) that this is not a recent development: For decades, the globally available food has been sufficient in principle to feed all people.

Thus, over the past 60 years, food production has grown even faster than world population, and this remarkable trend continues even today.

However, the food supply and the most important relevant structural data in the individual world regions differ very considerably (Table 7.2).

Table 7.1 Food per capita of the world population (statistically), 2019, in kcal/day

Food	kcal/day
Cereals	1310
Fruit and vegetables	200
Meat	240
Dairy products	180
Vegetable oils	300
Sugar	230
Other	500
Total	2960

Source: FAOSTAT

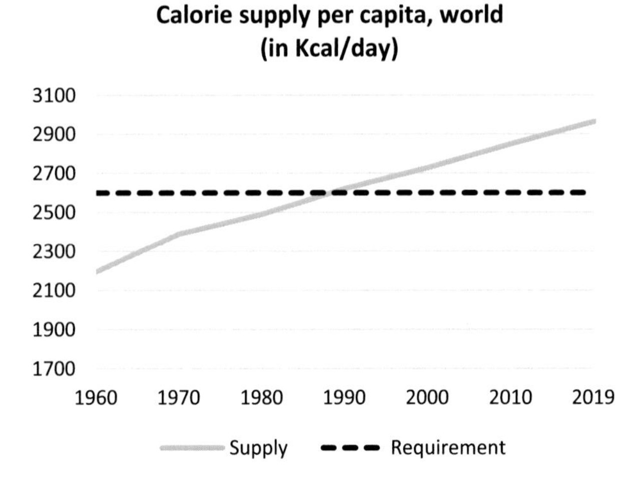

Fig. 7.1 Calorie supply per capita of world population (statistically), 1960–2019. (Source: FAOSTAT)

Table 7.2 Structural data on food supply in world regions, 2016

Region	Population (billion)	Cropland per capita (ha)	Specific grain yield (t/ha)	Food supply (kcal/day)	Malnutrition (%)
EU	0.5	0.22	5.5	3400	<2
North America	0.4	0.48	7.4	3650	<2
South America	0.4	0.38	5.0	3050	6
Asia (excl. SA)	2.6	0.13	4.9	3000	11
South Asia	1.9	0.13	3.2	2500	16
Africa	1.2	0.22	1.6	2600	19
World	7.4	0.22	4.1	2900	10

Sources: FAOSTAT, own calculations
SA South Asia

A number of essential conclusions can be drawn from this list:

- The main problems regarding malnutrition in the world are in South Asia and in Africa.

- In Asia, relatively little arable land per capita has been developed, partly because of the high population density. However, while most countries in Asia manage to compensate for this by high specific yields and, in some cases, by food imports, this is not always the case in South Asia.
- Africa has significantly more arable land per capita than Asia (about the same as the EU).[3] But due to very low yields per hectare, significant parts of the population in many countries are undersupplied (despite food imports averaging around 10% of the total supply).
- In North America, South America and Europe, there is sufficient arable land as well as high yields due to a more advanced agriculture. As a result, the population is generally well provided for.

Even within the regions of the world, there are considerable differences – not visible in Table 7.2 –, and problems of malnutrition are quite often caused by unfair or insufficient distribution of available food.

In the discussions about the global food situation, meat consumption is often the focus of attention. Indeed, more than 30 million km^2 of land is used for grazing, and another 5 million km^2 (= 30%) of the global cropland is used for animal feed. This means that 70% of the total agricultural area is used for livestock. On the other hand, animal products (meat, dairy products, eggs) contribute only just 20% to the food supply of the world's population. This is a clear disproportion, in a way, and perhaps you have encountered the following argument: If everyone kept to a vegetarian diet, there would be enough food in the world and no one would go hungry.

This book is not the place to really discuss this issue, but three aspects should be mentioned:

[3] Of course, arable land per capita varies considerably within Africa – but virtually all countries have at least 0.1 ha/capita available, about the same as the UK or Italy. The core problem, therefore, is low yields per hectare.

- The causes of malnutrition are certainly manifold and vary from region to region; but in any case, according to the figures presented above, the cause is *not* that too little food would be produced worldwide. Therefore, a vegetarian/low-meat diet for all is neither necessary for the elimination of hunger in the world, nor, conversely, would such a change guarantee an adequate food supply in all countries.
- The high meat consumption in Europe, North America and South America is not the cause of malnutrition in other parts of the world; rather, these three world regions export agricultural products to a considerable extent (see below).
- A lower-meat diet would nevertheless have tangible benefits: The pressure of agriculture on ecosystems generated by the drive for high yields could decrease somewhat, and agriculture's contribution to climate change would also decline. However, this latter effect is often overestimated: if global meat consumption were halved, greenhouse gas emissions would only fall by about 3%.[4]

Regardless of these considerations: people all over the world want to eat meat, dairy products and eggs (to varying degrees from region to region); and the cropland and pasture land that has now been developed – plus modern agricultural techniques – give them the opportunity to do so.

The Facts – The West

When it comes to nutrition, the average Westerner – i.e., one of the approximately 850 million inhabitants of the EU and the U.S. combined – differs from the average global citizen primarily in two respects (Fig. 7.2):

[4] In Germany, for example, total meat consumption causes only about as much greenhouse gas emissions as two large lignite-fired power plants.

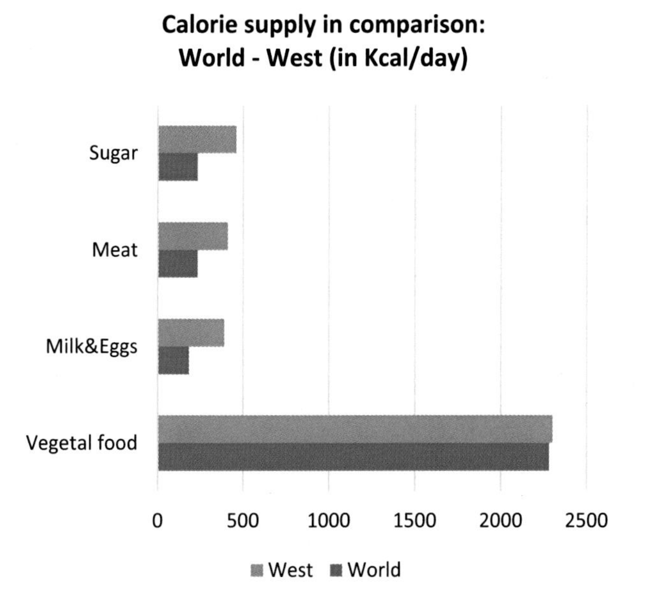

Fig. 7.2 Comparison of dietary pattern in the West (EU + USA) and in the world in kcal/day, 2019 data. (Source: FAOSTAT)

- The share of animal products is higher: 29% vs. 18%.[5]
- Westerners consume twice as much sugar: 460 kcal/day compared to 230 kcal/day.

The average person is actually oversupplied with these calorie numbers. Figure 7.3 shows that food consumption in the West has risen steadily over decades and has now almost stagnated at a high level for about 20 years.

There are a number of problematic aspects associated with these figures: Food wastage, widespread obesity, often questionable treatment of farm animals.

[5] The focus here is often on the high meat consumption in the West. And in fact, the Westerner covers on average about 12% of his caloric needs by meat, the world citizen 8%.

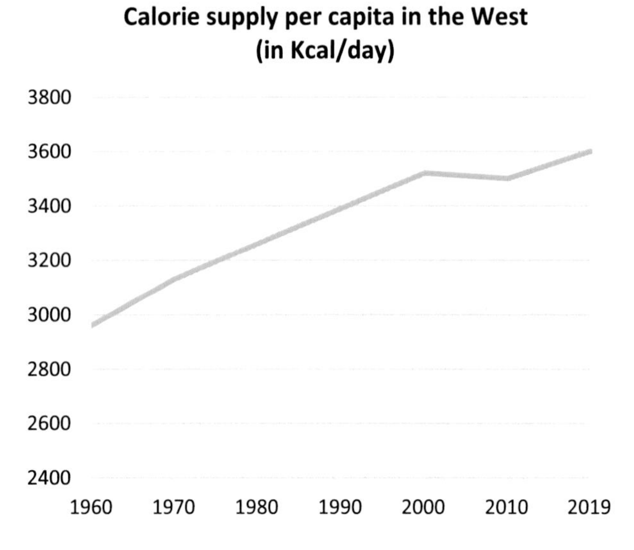

Fig. 7.3 Statistical calorie supply per capita in EU + USA, 1960–2019. (Source: FAOSTAT)

However, the assertion that this eating pattern goes at the expense of other regions and is only possible by exploiting resources elsewhere on earth – this assertion is essentially unfounded. On the contrary: The West *exports* (net) a considerable amount of food, especially cereals, feed (corn, soy)[6] and dairy products.[7]

[6] It is true that the EU imports (net) about 20 million tons of feed (mainly soy) from South America, and it thus (indirectly) uses about 5% of the arable land there (about 7 out of 130 million ha). But it could also get this from the USA. This means that, overall, the West is completely self-sufficient in feed and thus in meat production. By the way, the EU could also produce these feedstuffs itself, but then it would no longer be able to export grain and would have to stop the use of arable land for energy purposes (which does not make much sense anyway).

[7] Orders of magnitude of (net) exports in 2017 according to FAO: 100 million tons of grain, 35 million tons of feed, 30 million tons of dairy products, ten million tons of meat. In contrast, there are mainly about 30 million tons (net) imports of fruits and vegetables and ten million tons (net) imports of fish.

The reason for this can be seen in Table 7.2: The EU and the USA have both above-average cropland and a highly productive agriculture with top yields. Over the last 50 years, both together have led to an abundant, meat-intensive supply of the population, and they also allow a contribution to the nutrition of other world regions.[8]

Projection Until 2050

What will global food demand look like in 2050?

The FAO expects additional demand for arable products (vegetal food, feed) in the order of about 4 billion Gcal compared with today's levels[9] (see Fig. 7.4). It is assumed to be relatively certain that the (statistical) calorie supply per capita of the 10 billion people will be 3000–3100 kcal/capita. More difficult to forecast is the question of how eating habits will develop in the world, especially meat consumption.

The additional 4 billion Gcal can mainly be realized through yield increases on existing arable land (+3 billion Gcal); in addition, the FAO expects an increase in global cropland of 1 million km^2 (+1 billion Gcal).[10] Both forecasts seem quite realistic in light of Table 7.2[11] and the previous chapter (considerable potential of arable land).[12]

---

[8] The West (EU + USA) is the world's largest net food exporter, ahead of South America and Russia.

[9] Cf. FAO (2012, 2017).

[10] This expansion of arable land can even be dispensed with if energy production via agricultural products ceases, which I consider likely (cf. Chap. 9)

[11] For cereals, this requires an increase in yield on world average from 4.1 t/ha today to around 5 t/ha.

[12] Perhaps you are wondering whether artificial fertilizer production can keep pace with such a development. See Chap. 10 (Phosphorus).

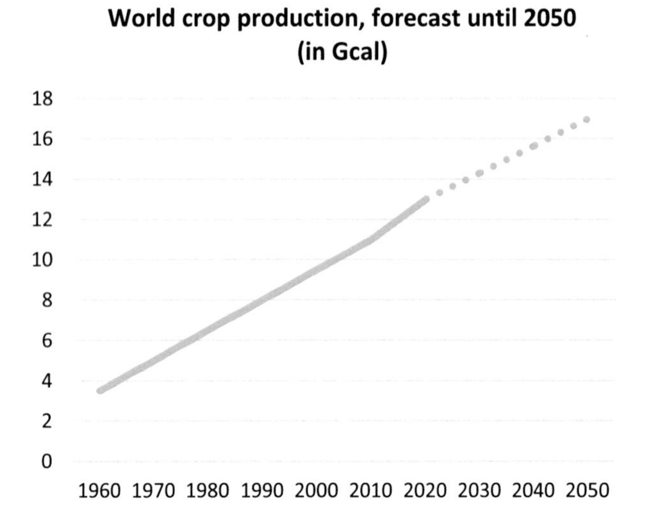

Fig. 7.4 Crop production of the world 1960–2050, from 2020 projection values. (Sources: FAO (2012), own calculations)

In other words, good nutrition for 10–11 billion people is clearly possible with a small expansion of cropland and with the agricultural techniques already available today.

Evaluation/Summary

Over the last 50 years, the world population has grown faster than ever before. During the same period, global agriculture has accomplished an amazing feat that many observers did not believe possible in 1970: It has increased food production even faster. This was made possible on the one hand by a moderate expansion of cropland (+15%), but above all by a drastic increase in the use of artificial fertilizers (+600%).[13,14]

However, the performance of agriculture varies greatly from one region of the world to another (and in some cases also from country to country). Indeed, the cause of malnutrition in the world – which still affects 800 million people – is not a lack of arable land, i.e. not a lack of natural resources. Rather, the core problem is the lack of effective use of available resources in individual regions, especially in parts of Africa and South Asia.[15] To put it positively: If agriculture in the countries concerned could be developed in a similar way to that in the EU (as has already happened in other regions of the world) and thus achieve cereal yields of 4–5 t/ha, hunger will have been largely defeated there.

In other words: Yes, goals 1 and 2 of the "Agenda 2030" can be achieved. And with our agricultural ecosystem of 10% of the earth's surface, we can also feed 10 or 11 billion people well and healthily in the long term.[16]

Excursion: Soil Degradation

One of the greatest dangers to the food supply of mankind, it is often said, arises from the gradual deterioration of arable soils ("soil degradation") with potentially drastic consequences for future crop yields.

---

[13] In 1960, about 33 million t of artificial fertilizers were used worldwide; today the figure is over 200 million t.

[14] It is also noteworthy that over longer periods of time – i.e., apart from short-term fluctuations – world market prices for food are fairly constant, cf. Fig. 10.12.

[15] The broader question of what, in turn, is the cause of this state of affairs suggests itself, but it goes beyond the scope of this book.

[16] This is especially true when one considers the new agricultural techniques under development around the globe: Hydroponics in Japan and China, fabric net greenhouses in India, brackish water cooled greenhouses on the edge of the Sahara, artificial meat production in the U.S., and many others.

The most important and critical problem here is *soil erosion*, i.e. the removal of upper soil layers by water and wind. Soil erosion is actually a natural process, and on about 85% of the earth's land areas (forests, savannahs, pastures) it is completely uncritical, with average values of about 1 mm of soil loss per decade. On our cropland, however, the rates of loss are on average almost ten times higher – simply due to the particular form of use "arable farming."

However, the values vary greatly around the world and depend on the type of soil, the climate zone, the slope of the cropland[17] and, above all, the agricultural methods used. Figure 7.5 provides an overview of the degree of soil erosion on global cropland.

1. Soils with moderate erosion (<10 t/ha/year) lose on average about 0.5 cm of soil per decade; this level of erosion

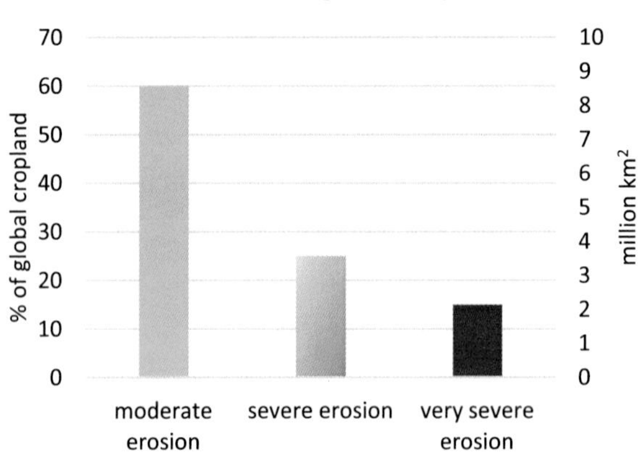

Fig. 7.5 Soil erosion on the world's croplands, 2012 data. (Sources: Borrelli et al. 2017, own calculations)

[17] Obviously, steeply sloping land is much more at risk than flat land. Terrace cultivation provides a significant remedy.

can be offset by higher fertilizer use, even in the longer term.

2. Soils with severe erosion (10–20 t/ha/year) lose an average of about 1 cm of soil per decade. Here, unless the agricultural methods used are improved, crop losses must be expected in the longer term. The extent of such losses in turn depends on climate, crop grown, soil composition, etc., but is estimated to be in the order of 1–3% per decade, i.e. 3–10% by 2050.
3. Very severe soil erosion (>20 t/ha/year) affects 2–2.5 million km^2 (out of 16 million km^2) of cropland worldwide. These soils are expected to be lost in the next decades, i.e. they will have to be abandoned for crop farming. Figure 7.6 shows the extent to which the individual world regions are affected. In the recent past, arable land has already been lost at a rate of about 0.5 million km^2

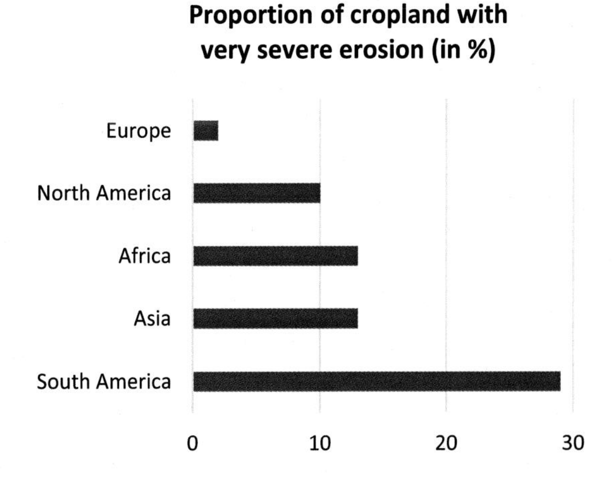

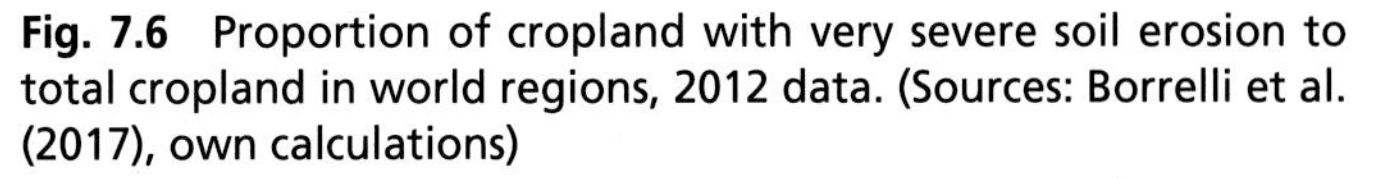

Fig. 7.6 Proportion of cropland with very severe soil erosion to total cropland in world regions, 2012 data. (Sources: Borrelli et al. (2017), own calculations)

per decade; thus, a similar rate can be expected for the foreseeable future.

What do these figures mean for the food situation in the twenty-first century?

First of all, it should be emphasized that the problem of soil erosion can be greatly reduced by improved agricultural methods – suitable crop rotations, no plowing before sowing, terraced cultivation, and many others. As awareness of the problem has increased in recent years, significant progress can be expected in the future.[18] But if we disregard such improvements, the following picture emerges:

- Soil losses (2) entail only a 1–3% decrease in total world crop yields in 2050; this is of little significance given the expected yield increases from better agricultural practices and fertilizers (see main text).
- The loss of soils (3) can be compensated – as in the past – by the development of new arable land, since sufficient suitable land is available (cf. Chap. 6).

Conclusion Soil erosion is a serious problem on the world's arable land. On the one hand, however, it can be greatly reduced by already known agricultural techniques, and on the other hand, it does not endanger the food supply of mankind, at least not in this century.

[18] However, the willingness of local farmers to change is limited by the fact that the effects of even severe soil erosion (i.e., the resulting yield losses) are relatively small and therefore only noticeable over decades.

8

Drinking Water

Introduction

"The global drinking water crisis threatens our prosperity," headlined *Die Welt* in August 2019. "3.6 billion people affected by water shortage," wrote *Die Zeit* in March 2018. "The world is facing a historic drinking water shortage," read *Stern* a few years ago.[1] Or even: "The planet is dying of thirst," as the well-known scientist Harald Lesch put it in 2016.[2] Concerns about water scarcity are actually much older: "The next war in the Middle East will be fought over water," predicted the then UN Secretary General Boutros-Ghali in 1985.

Indeed: water is the basis of life par excellence. Humans can go 2 months without food, but only 3–4 days without drinking water. Therefore, the question of whether or not there will be enough drinking water for 10–11 billion people in the future is at least as important as the corresponding question in the previous chapter regarding food.

Boutros-Ghali's prediction has not come true, neither for the Middle East nor for other regions. In this chapter, we

[1] Welt of 8/21/2019, ZEIT of 3/19/2018, Stern of 3/20/2015.

[2] Headline of a chapter in Lesch (2016).

© Springer-Verlag GmbH Germany, part of Springer Nature 2022

T. Unnerstall, *Factfulness Sustainability*,
https://doi.org/10.1007/978-3-662-65558-0_8

will see that the other aforementioned headlines should also be taken with great caution.

The Facts – Global View

The earth is big, and there is quite a bit of water on it. More precisely, 1.4 billion cubic kilometers (km^3), or 1.4 quadrillion liters. Now, over 99% of this is salt water or ice; but even the 1% liquid fresh water is still impressively large at 13 million km^3 or 1.7 billion liters per Earth citizen.

Again, about 1% of this amount – a little over 100,000 km^3/year – circulates. It falls as precipitation on the land[3]; about 60% evaporates (i.e. goes directly back into the cycle), 40% seeps into the ground and then returns to the cycle via groundwater, rivers, lakes and finally the sea. These latter water amounts, about 40,000 km^3, are therefore often referred to as "renewable water resources". They are newly available to terrestrial ecosystems every year.

And what do humans do? Mankind withdraws a relatively small amount – about 4000 km^3 – from these renewable water resources (i.e., from groundwater, rivers, and lakes) each year. It uses this amount for various purposes – agriculture (70%), industry (18%), households (12%) – and then returns it to the cycle. We thus currently use only one tenth of the renewable water resources. Historically (Fig. 8.1), water use increased sharply in the second half of the twentieth century, but has been growing only slightly for the past 20 years.

These few figures show: There is no global shortage of drinking water, and there is certainly no question of a "planet dying of thirst." On the contrary, there is an

[3] What is meant here is the solid surface of the Earth without ice, i.e., about 130 million km^2.

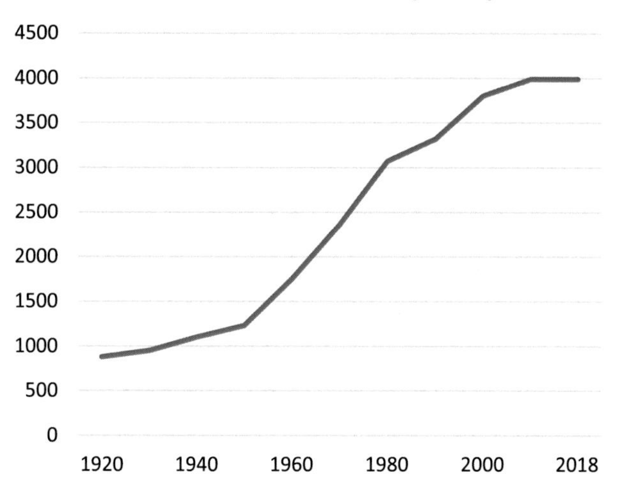

Fig. 8.1 Water withdrawal of the world, 1920–2018. (Sources: OurWorldinData (Water Use and Stress), UN 2019, FAO 2021)

abundance of water and freshwater on this planet, and humans use only a very small part of it.

But how do such headlines as quoted in the introduction come about? Well, there are four aspects:

1. Globally, (fresh) water is abundant, but it is quite unevenly distributed, and in a few regions/countries, natural freshwater resources are indeed scarce or insufficient.
2. In a number of other countries, water is in principle abundant, but it is unevenly distributed spatially and/or temporally throughout the year, so that – without compensatory technical measures – shortages can occur in individual parts of the country and or in individual months (and possibly flooding in others).
3. After human use, the extracted water is returned to the cycle, but it often contains additional substances –

mainly chemicals (households, industry) or fertilizers (agriculture). This problem of *water pollution* has nothing to do with *water scarcity*, but is often discussed in the same context.

4. In a number of countries around the world, there is actually always enough fresh water, but due to a lack of even basic technical infrastructure, many people there still do not have access to (clean) drinking water. This concerns – similar to the problem of malnutrition or lack of agricultural development – mainly sub-Saharan Africa and South Asia. Since this is actually a different problem, namely an issue of socio-economic development,[4] we will not discuss it in detail below.

So let's take a closer look at aspects (1) through (3).[5]

Re 1)

90% of all countries on Earth (i.e., about 200 out of 220) have more than enough fresh water available – on a nationwide and annual basis. They use less than half of the (annually newly formed, i.e.) renewable water resources, the vast majority even less than a quarter. Only in about 20 countries (with a combined population of about 700 million) are natural fresh water resources actually scarce or simply insufficient to meet demand. These countries are mainly concentrated in two regions: North Africa and the Middle East, including the countries bordering them in Asia.

[4] Significant progress has been made in this area over the past 30 years: both the number of people without access to basic water treatment has been cut in half (to about 600 million now) and deaths due to unclean drinking water have been reduced.

[5] The following data are from FAO's AQUASTAT database and the World Atlas of Desertification (EU 2018).

For the majority of these countries, the technology of seawater desalination plants offers the possibility to supplement the drinking water supply or to make it possible in the first place – they are located by the sea and thus have access to de facto unlimited water resources. To this end, they do have to make above-average investments in water infrastructure.[6] In other words, these countries have a real problem, but it can be solved permanently with existing technologies.[7]

The situation is much more difficult and potentially conflict-prone for countries whose water supplies are highly dependent on transboundary rivers and whose needs (especially for agriculture) can therefore not be met by their internal water resources. These are primarily Egypt, Jordan, Iraq, Syria, Pakistan, Turkmenistan and Uzbekistan.[8]

Re 2)

10–15 countries around the world have sufficient water resources available throughout the country and over the year, but they are confronted with water shortages in individual parts of the country and/or at individual times of the year. Specifically, this applies to[9] Northwest India, northern and

---

[6] However, in the future energy world, which is mainly based on solar and wind energy (cf. Chap. 9), they will have to pay *below average* for energy supply: Without exception, these countries are blessed with a lot of sunshine and, in most cases, a lot of wind.

[7] A good example of this is Israel, cf. Global Voices (2016).

[8] Probably the greatest difficulties currently exist in Turkmenistan and Uzbekistan: cotton cultivation, already established in Soviet times, was only possible with massive artificial irrigation from the adjacent rivers. This (over)use, which is ultimately the result of a political mistake, has led in particular to the extensive drying up of the Aral Sea, with devastating effects on the environment and on the living conditions of the local population.

[9] Cf. World Atlas of Desertification (EU 2018), Chapter "Water Stress and Urbanization."

western China, parts of southern Europe (especially southern Spain), Ukraine, eastern Turkey, Afghanistan, southern Australia, parts of Argentina and Chile, northern Mexico, the southwestern United States and parts of South Africa. Because the solution to this problem has often been to pump up more groundwater than is newly formed, many of these regions are seeing a drop in groundwater levels (and in some cases, a slow drying up of overexploited lakes).

In such cases, one or more of the following solution strategies may be considered:

- Relocation of particularly water-intensive agriculture (by far the largest water consumer) to other parts of the country
- Water storage/transport pipelines for temporal/regional compensation
- Use of seawater (again, most affected areas are located at or near the sea)

Thus, this form of water scarcity is not an inevitable fate or an unstoppable, dangerous crisis: what is needed is long-term policy, combined with modern technical infrastructure.

Re 3)

There are two quite different sets of problems with water pollution. Wastewater from *households and industry* can and should be treated in sewage treatment plants – this is state of the art, but by no means standard practice in all countries of the world. Similar to the shortage of drinking water, however, this is essentially an issue limited to individual regions.

More serious and globally relevant – in the sense that almost all countries in the world are confronted with it in one way or another and need to develop solutions – is the problem of *agricultural* wastewater. The main issue here is surplus fertilizers that end up in groundwater, rivers and lakes, and ultimately in the oceans. This topic is the subject of Chap. 15.

The Facts – The West

The West – the EU and the USA considered together – has renewable water resources similar to the rest of the world: on average, 5200 m^3 (5.2 million liters) per inhabitant per year. Of this, only a little over 10% is used, i.e. the annual per capita water withdrawal amounts to 600–700 m^3. It is noteworthy that water consumption in the West has been declining for decades (see Fig. 8.2). Thus, overall, there are

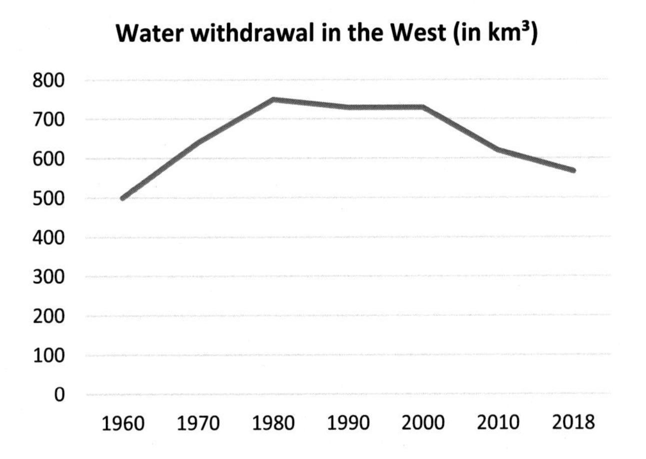

Fig. 8.2 Water withdrawal by USA + EU, 1960–2018. (Sources: OECD, USGS, OurWorldinData (Water Use and Stress), FAO 2021, own calculations)

no real water scarcities here; only in the southwest of the USA (incl. California) and in southern Spain are there regional problems that can be solved by appropriate technical infrastructure.

So, rather clear skies on this issue – but there is one aspect that seems to cloud the picture: the accusation that the West contributes to fresh water problems in other countries of the world through so-called **virtual water imports**. What does this mean? Each traded product – in particular, those agricultural products that require large amounts of water for artificial irrigation – can be assigned a certain amount of water that is needed on average for its production. Thus, when a country imports a certain product, it is indirectly responsible for a certain amount of water consumption in the country of origin. Since this amount of water is no longer available there, it is, so to speak, "virtually" exported or imported along with the product. Each *real* export-import balance of products of a country corresponds in this way to a *virtual* export-import balance concerning water.

In 2011, the UN commissioned two Dutch scientists to painstakingly compile these data for all countries.[10] The main result for the West is that its virtual water balance is indeed negative. As we have seen in Chap. 7, the West does export agricultural products in total, but it produces them as a rule without artificial irrigation (i.e., without water consumption in the sense discussed here), while a number of imported agricultural products are produced with artificial irrigation in the countries of origin and are therefore water-consumption-intensive. However, the size of these virtual net imports is very moderate: it is less than 100 km^3

[10] Hoekstra and Mekonnen (2012).

and thus in the order of only 2–3% of global water consumption.[11]

More importantly still, such a negative virtual water balance per se is quite unproblematic – as long as the main virtual water imports come from countries with sufficient water resources. It is an essential characteristic of reasonable world trade that countries primarily export those products for the production of which they possess the necessary natural resources. And indeed, in general this is precisely the case.[12] The 20 countries with a real scarcity of fresh water export – not surprisingly – hardly any agricultural products anyway.

For these reasons, the accusation that the West is partly responsible for the problems in fresh water scarcity regions is essentially untenable.[13]

Projection Until 2050

To forecast the future development of global water consumption, it makes sense to look at the foreseeable per capita consumption (and then multiply it by 10–11 billion people). Figure 8.3 shows the clearly decreasing tendency of

---

[11] We consider here only water consumption in the sense discussed so far, i.e., the abstraction of water from groundwater, rivers, and lakes. In addition, the UN study also surveys the use of so-called "green water" – i.e., natural rainfall – for the production of agricultural products. However, since this does not represent any human intervention in the water cycle and therefore has no influence on the drinking water situation in the countries, this part of the virtual water balance is irrelevant for the accusation discussed here.

[12] The worldwide trade relations concerning agricultural products are documented in the database FAOSTAT (Trade).

[13] Of course, this does not exclude exceptions; in individual cases, imports may also originate from regions with (seasonal) water shortages. In particular, cotton from Turkmenistan and Uzbekistan should be mentioned here, compare footnote 8.

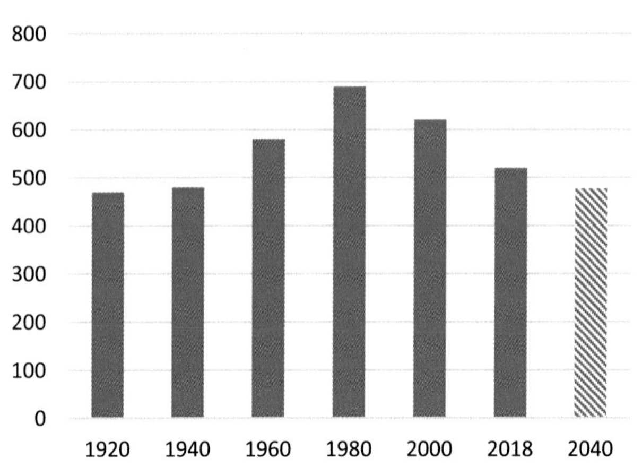

Fig. 8.3 Water withdrawal per capita, world, 2040 = forecast value. (Sources: OurWorldinData (Water Use and Stress), UN 2019)

this parameter, so that one can assume a global average of 450–500 m^3/capita for the middle of the century. It is therefore easy to understand why the UN currently forecasts a water demand of about 5000 km^3 for the year 2050.[14] Water consumption would thus reach an order of magnitude of 12% of renewable resources (= 5% of the natural fresh water cycle).

[14] UN Water Development Report (UN 2019). Even using the West's current per capita consumption of 600-700 m^3, global water use would increase to a maximum of (700 m^3 × 11 billion people ≈) 8000 km^3 or (only) 20% of renewable resources.

Evaluation/Summary

Water is the basis and elixir of all life,[15] and this also applies to us humans. All historical settlements were built by our ancestors on rivers and lakes, and even today most of the world's major cities are located close to water. The challenge of "fresh water supply" has been successfully solved on this basis in all regions throughout human history.

On sober analysis, all available data show that, in principle, we need not worry about the water supply of mankind in the future. Even with further intensification of agriculture towards more artificial irrigation and even with increasing personal water consumption of 10–11 billion people, humanity's water use will be well below 20% of the globally available renewable water resources. Put another way: Mankind only has a minimal impact on the Earth's natural water cycles in terms of water quantity.[16]

However, this natural water cycle functions very differently from region to region. If man decides to settle and farm in regions where fresh water is not available in sufficient quantities for his purposes – either throughout the year or at least during individual periods –, he must compensate for this with technology.[17] Such technologies have been available for a long time; they are just not (yet) used to a sufficient extent in all affected regions: be it seawater desalination plants (especially for the 20–25 or so countries that have too few natural freshwater resources overall or in

[15] The primary method for exploring extraterrestrial life is to search for water on Earth-like planets outside the solar system.

[16] In terms of water quality, the picture is somewhat different. The main problem in this respect is the considerable quantities of (excess) fertilizers that enter the lakes and seas via the water cycles; cf. Chap. 15.

[17] The same is true in principle for temperature: humans can survive in numerous regions of the Earth only because they use heating techniques to compensate for the ambient temperature.

large parts of the country), water transport pipelines and water storage facilities (especially for the 10–15 countries with mainly seasonal shortages); or, of course, basic water infrastructure (for countries in which people have no access to clean drinking water despite abundant water resources).

Characterizing this situation with the term "global drinking water shortage" is quite common and leads to the headlines quoted at the beginning, but it conveys an essentially false picture. Also, to suggest a supposed responsibility of the West for regional water problems elsewhere misses the point – the much cited "virtual water imports" of the West do not really matter on a global scale, nor do they matter for most virtual water exporters.

With this conclusion, I am far from wanting to play down the very real and quite serious problems of scarce water supply in a number of countries. I just want to make this clear: As with food supply, these problems do not stem from a lack of natural resources, with a few exceptions. Rather, the existing regional or local bottlenecks can usually be overcome with modern technology, without putting undue strain on ecosystems.

Of course, the changes to be expected as a result of (even moderate) climate change must be taken into account here. The relevant climate forecasts largely agree that there will be only minor changes in the *annual* rainfall in most regions of the world, but *seasonal fluctuations* will increase almost everywhere. In concrete terms, droughts and floods will become more frequent. To the extent that climate change is not prevented, additional water infrastructure is therefore one of the most important adaptation measures[18] – even in countries where drinking water out of the tap is the most natural thing in the world.

[18] … and it is easy to calculate that this will be much more costly and time-consuming than taking the necessary measures to mitigate climate change now. Why does it take so long for such a simple truth to translate into political action?

Part III

Energy and Raw Materials

9
Energy

Introduction

Humanity's current way of life and economy are often characterized as being unsustainable; and in order to underpin such assertions, the one parameter most often cited is probably energy consumption. Indeed, global energy consumption has exploded over the last 100 years and is now almost nine times higher than it was in 1920, while the world's population has grown by a factor of only four. Around 85% of this energy consumption comes from *fossil fuels* (i.e., energy sources that were created in previous eras and stored in the earth's crust): Coal, oil and natural gas. The historical development is shown in Fig. 9.1.

This status quo is characterized as unsustainable in two respects:

- First, fossil fuels are obviously *non-renewable resources*. They are finite, and for many years, experts have repeatedly issued warnings of "peak oil": the point in time at which oil production will start to decline inexorably due to the depletion of the Earth's reserves – while global demand remains constant.

© Springer-Verlag GmbH Germany, part of Springer Nature 2022

T. Unnerstall, *Factfulness Sustainability*,
https://doi.org/10.1007/978-3-662-65558-0_9

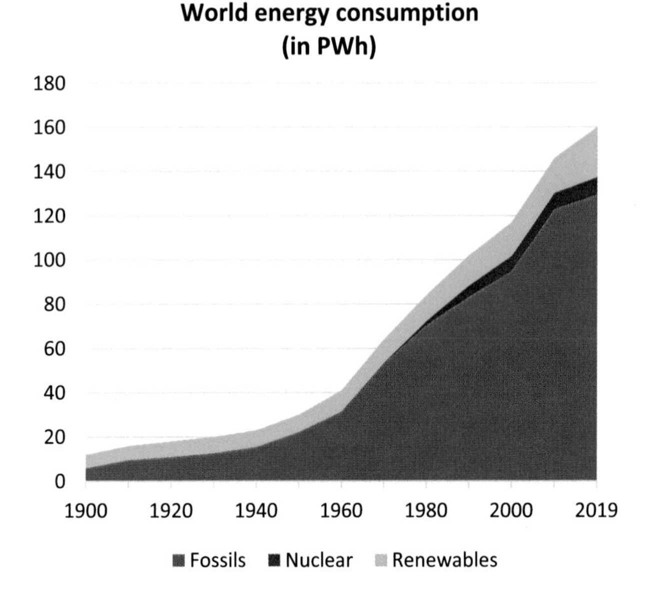

Fig. 9.1 Primary energy consumption of the world, 1900–2019, in PWh (PWh = 1 trillion kilowatt-hours). (Source: OurWorldinData (Energy))

- On the other hand, the use of fossil fuels is by far the most important *cause of climate change* – around 75% of global greenhouse gas emissions are attributable to the burning of coal, natural gas and oil. Conversely, it is clear that if mankind wants to curb climate change to such an extent that the consequences for ecosystems and, in particular, for its own living conditions remain reasonably foreseeable and manageable, it must largely end the use of fossil fuels in the next 30–40 years.

But is that even possible? How can global energy consumption be covered in the future – especially if it could once again be much higher than today due to the further (albeit limited) increase in the world's population and due to further significant increases in living standards in most countries?

The Facts/Projection Until 2050 – Global View

A very rough calculation – assuming 10–11 billion people and a per-capita energy consumption as high as in the West today – would yield a forecast of around 500 PWh/year for the second half of this century. However, the actual current predictions regarding the future energy consumption of mankind look different: They expect a maximum of annual world energy consumption around mid-century in the order of 200, max. 250 PWh (Fig. 9.2).

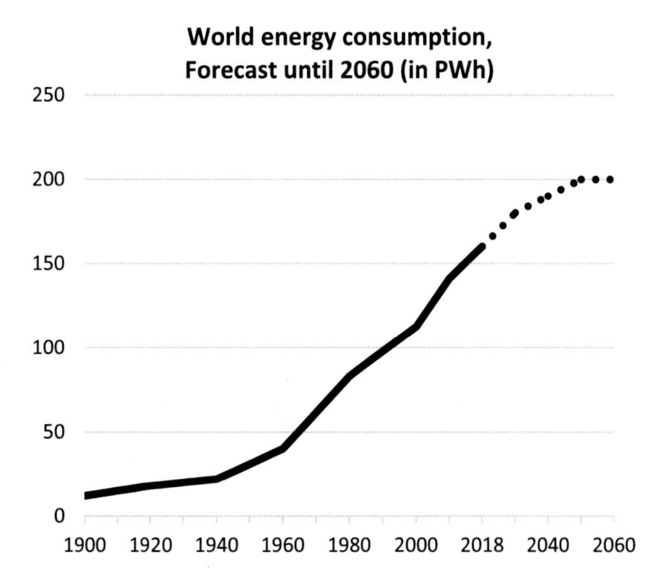

Fig. 9.2 Primary energy consumption of the world 1900–2060, from 2018 projection, in PWh. (Sources: McKinsey 2019, BP 2019, IEA 2019)

The difference is due to the fact that very significant increases in energy efficiency can be expected, both in the use of energy for transport, heating, industrial production, etc., and in the upstream conversion processes (i.e., mainly electricity production).[1] 200 PWh/year – so how to provide such an amount of energy?

In fact, in principle, this is not a problem at all, as the following three considerations show.

Fossil Energies

We know today that there were about 170,000 PWh of fossil energy reserves in 1850 – that is, before humans began to extract and burn first coal, then oil and natural gas on a large scale. Of this, to date humanity has consumed less than 4%! These reserves would still last for more than 800 years to supply the 200 PWh/year (not counting the reserves not yet discovered today).

However, this path is blocked due to the associated CO_2 emissions. Therefore, the fossil age, which started around 1850, will turn out to be only a short episode in human history of about 200 years. And in the end, 95% of the fossil fuels originally available will have to remain in the ground.

Renewable Energies

The earth's surface receives 750,000 PWh(!) of solar energy every year. Restricting the consideration to the easily accessible land areas with only sparse vegetation (thus leaving out forests, arable land, mountains, ice surfaces, etc.), then

[1] Between 1960 and today, the energy efficiency of the global economy has doubled – and the expectation is realistic that it will double again (at least) in the next 50 years.

conservatively calculated it is still about 60,000 PWh/year.[2] This energy can be converted with modern PV modules into electricity and then, partly, into hydrogen and so-called synthetic fuels.[3]

In this way alone, there is a theoretical potential of available renewable energy sources in the order of 5000 PWh/year. Ad to this the potential for wind energy[4] of over 1000 PWh/year. In other words: Mankind can produce CO_2-neutral, renewable (i.e. sustainable) energy in abundance – with technologies that are already available today.[5]

To put it the other way around: in order to provide 200 PWh of climate-friendly energy per year, using the technologies mentioned above, it would take[6] about 1.5 million km^2 (= 0.3% of the earth's surface). Considering that, for example, more than 15 million km^2 of otherwise hardly usable deserts are available for such a purpose,[7] we can draw the following conclusion: At least from the energy side, there are de facto no limits to the further development of mankind.

---

[2] We calculate here with 30–40 million km^2 area with a solar radiation of 1500–2000 kWh/m^2.

[3] Synthetic fuels are liquid or gaseous hydrocarbons that are chemically largely identical to petroleum-derived gasoline/kerosene or natural gas, but are CO_2-neutral: They are synthesized in large-scale plants from hydrogen (obtained via electrolysis mainly from PV or wind power) and CO_2, and when burned emit as much CO_2 as was used to produce them.

[4] According to a study published in 2019, there is a potential of about 420 PWh/year for offshore wind farms alone.

[5] 6000 PWh/year is about 30 times the projected annual energy consumption of humanity in this century.

[6] You may be wondering whether there are enough raw materials for the solar panels to support such a dramatic expansion in the use of photovoltaics. Well, the main raw material is silicon, and according to today's technology, about 2 Gt of silicon is needed to produce the necessary approx. 200 TW of PV modules. However, the natural deposit in the earth's crust of silicon is 100 million Gt.

[7] The transport from the centers of production of hydrogen/synthetic fuels to all parts of the world can – like today's transport of coal, crude oil and natural gas – be carried out partly via pipelines and mainly via shipping. The existing infrastructure can be used to a considerable extent for this purpose.

Nuclear Fusion

There can be no serious doubt that mankind will eventually master the technology of nuclear fusion (and thus directly tap the energy source of the sun, so to speak) – probably already in this century,[8] but if not, then with some certainty in the twenty-second century. Since neither the safety nor the final storage problems of conventional nuclear power plants are a factor here, and since no CO_2 is emitted, this would be a second permanent technological solution for the future energy supply of mankind.

In principle, the necessary fuels (hydrogen isotopes) are finite resources. However, the earth's energy supply is de facto unlimited: for the currently predicted and every conceivable energy demand of mankind, it would last for millions of years.

Alas, for the climate protection goals humanity must achieve over the next decades, this technology comes too late.

"The technological feasibility may be there, but what about the cost?" you may now be thinking – and the question is very valid. Energy is a fundamental, indispensable part of everyday life: heating, household appliances, communication, mobility; it also forms the basis of all economic activity and industrial production. Energy is therefore one of the decisive keys to societal development toward higher standards of living. Thus, the affordability of energy sources is of central importance. Everywhere in the world, cost plays the most important role when it comes to the necessary "*energy transition*" (i.e., the transition from the fossil energy world to the world of renewable energies).

[8] Scientists working on current nuclear fusion projects assume that they will be able to build the first commercially viable power plant as early as around 2050.

So what is the cost of an energy supply that relies primarily on PV and wind power instead of coal, oil and natural gas[9]?

This question is so complex and so different to answer for different regions of the world that it would justify a book of its own. However, the main trends can be estimated relatively well:

- The cost of PV and wind power plants (including the necessary storage and conventional backup power plants) are already in a similar range to most fossil fuel power plants, and costs continue to fall.[10]
- E-mobility in the passenger car sector will almost certainly soon be cheaper than gasoline/diesel mobility, because e-cars are easier to build and require less maintenance, because they need much less energy (electric motors are more efficient than internal combustion engines by a factor of three to four), and because the essential cost factor "battery" will become less important due to the current construction of huge battery factories.
- The cost projections for hydrogen and synthetic fuels produced with PV electricity for 2050 are constantly being revised downward. Nevertheless, from today's perspective, these renewables will still be more expensive by a factor of two to three than the average world market price for the oil/natural gas mix in recent decades. In re-

[9] In addition to renewable energies, conventional nuclear power can provide part of the solution in an number of countries. However, such power plants are considerably more expensive today than a PV/wind-system (incl. Storage capacities etc.), and they have other drawbacks. New developments are underway, but all available studies to date estimate the contribution of (conventional) nuclear energy in an CO2-free world in 2050/2060 to be below 10%.

[10] In Germany, newly built large PV plants supply electricity for well under 5 ct/kWh. The latest large-scale plants in the Middle East even produce PV electricity for less than 1.5 ct/kWh. By comparison, the fuel costs of a fossil fuel power plant alone have averaged 2.5–4 ct/kWh over the past decade.

turn, however, the world economy is also expected to grow very significantly until 2050.[11] Hence, the *ratio* of fuel prices to household income/industry sales will probably hardly change.

Overall, in 2050 the global ratio "primary energy cost/economic output" is likely to be within the long-term corridor of 3–5%. In other words: energy will remain affordable even after the transition from fossil fuels to PV, wind and in part nuclear energy.[12]

Ultimately, however, the cost issue is actually secondary. Even if the above estimates turn out to be too optimistic, at least one truth can be considered certain: No matter how expensive this transition to the post-fossil age will turn out to be – it is in any case much, much cheaper than if mankind would have to pay for the consequences of a massive climate change (i.e. of frequent droughts, floods, tornadoes, hurricanes, heat waves with large-scale fires, etc.). In other words, there is simply no reasonable alternative.[13]

Therefore, we will have a completely different energy supply worldwide in 2100 – the only question is how fast in this century the technological change will happen, and therefore how successful we will be in mitigating climate change.

[11] In the last 30 years (1990–2020), global economic output has grown by an average factor of 2.2 per year.

[12] For Germany, this result has been confirmed in detail in several studies. There, the effects of an ambitious climate policy on prosperity and jobs for 2050 have also been estimated. The result: The overall impact is small – other factors will play a much more important role. See, e.g.,Federation of German Industries (BDI). ("Climate Paths for Germany," 2018).

[13] For the next decades (i.e. for a transitional period), an important alternative can be and probably will be the socalled CCS/CCU-technology. Here, the CO_2 is filtered out of the exhaust gases of fossil power plants, and then partly stored underground, partly reused. It doesn't look like this technology can compete cost-wise with renewables in the long run – especially not in many countries in Asia and Africa that have excellent conditions for PV power.

The Facts/Projection Until 2050 – The West

The West – the USA and the EU – consumes significantly more energy than the rest of the world: on average about 50,000 kWh per inhabitant per year compared to 20,000 kWh worldwide[14] (Fig. 9.3). The figure shows a clear downward trend.

In absolute terms, too, the West has already passed the peak of its energy consumption (Fig. 9.4). It will certainly continue to fall, but the exact development up to 2050 will

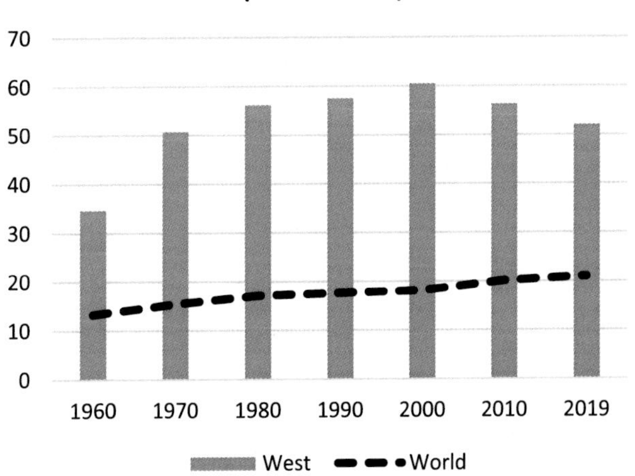

Fig. 9.3 Annual primary energy consumption per capita in the West (US + EU) and the world, 1960–2019, in 1000 kWh. (Source: OurWorldinData (Energy))

[14] Much of this energy has been imported in recent decades, but only for economic reasons: In principle, the Western countries' own fossil deposits would be quite sufficient to supply themselves with energy.

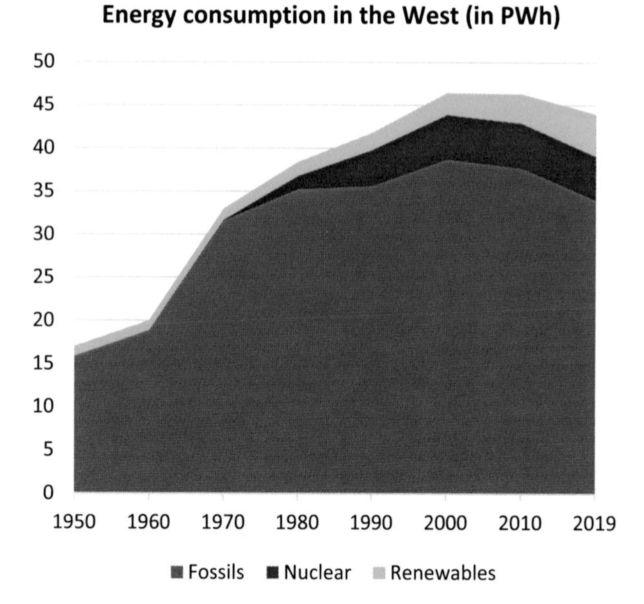

Fig. 9.4 Primary energy consumption of the West (USA + EU), 1950–2019. (Sources: ESTAT, IEA, OECD)

depend primarily on how quickly the switch in power generation (from coal and gas to renewable energies) and in transport (from gasoline/diesel to e-mobility) takes place.[15] Assuming an ambitious climate protection path, energy consumption in 2050 will be at least 20–30% lower than today.

At present, however, CO_2 emissions per capita are far above average due to the high energy consumption: ca. 10 t/year compared to a global figure of around 4 t (see Fig. 3.2).

[15] In these two sectors, the energy yield from fossil fuels is particularly poor: in electricity generation, it is 40-50%; in automobiles, it is only 20-30%.

History weighs even more heavily. True, the US + EU share of global CO_2 emissions is only about 25% *today*; but *historically*, the US + EU have been responsible for nearly half of the 1.7 trillion tons of CO_2 emissions that humanity has produced since the Industrial Revolution. For these reasons, there is – this seems to me indisputable – a moral obligation of the West to not only reduce its own CO_2 emissions particularly fast, but also to support other countries financially and technologically in their respective "energy transitions".

What is true for the world as a whole is also true for the West: It would be no problem at all for the USA and the EU to supply themselves self-sufficiently with renewable energies. In Germany, for example, about 7% of the country's surface area would be needed to provide the necessary regenerative energy sources (electricity, hydrogen, synthetic fuels) via PV-power. This is no more area than is currently needed for energy crops – a technology that has no future anyway, neither in terms of cost nor eco-balance.

However, at least the EU will probably choose a different path. In recent decades, importing significant quantities of fossil fuels from other countries as part of global trade has made economic sense, and it is likely to be the same with renewable fuels. Many regions of the world have much better natural conditions for the production of hydrogen and synthetic fuels than Europe because of their abundance of sunshine and wind, and thus importing energy is likely to continue to play a major role for the EU in the future (see also Excursion).

Evaluation/Summary

Mankind is consuming more energy than ever before. It is often claimed that this energy consumption – or in any case the energy consumption in the West – is "unsustainable" (and for this reason the consumption patterns causing it and the underlying economic system are said to be not sustainable either). This assertion is false.

Global energy consumption or the energy consumption of the West *as such* are no problem at all. The earth provides many times more energy resources than mankind can ever consume. What is not sustainable is the *particular way* mankind has covered its energy consumption in the last 150 years, namely with fossil fuels. And even this statement is not so much related to the finite nature of these resources, but rather to the CO_2 emissions associated with the use of fossil fuels.

In a nutshell, humanity does not have an energy problem, it has a CO_2 problem.

Switching from fossil fuels to renewable or nuclear, i.e. CO_2-free energy sources is technically not a real problem with technologies already available today. Moreover, the estimates available today show that it is also quite affordable in financial terms. And even if this transition were to cost a few percent in economic growth, or even if energy were to become a little more expensive in the future: The quality of life for future generations must be worth it.

Indeed: in the end, climate protection is not a question of technology or of finances. It is solely a question of the common will of the vast majority of countries. The Paris climate agreement of 2015 was a very significant milestone in terms of this common will. But today, 7 years later, global CO_2 emissions continue to rise, and in its concrete actions, humanity is still far from the path it prescribed for itself in Paris. The West in particular must finally assume its responsibility here.

Excursion: Saudi Arabia Without Oil Exports?

Saudi Arabia is the second largest oil producer in the world, behind the USA. Around 600 million tons (= 7 PWh) of crude oil are extracted from the earth there every year – that is 4% of mankind's energy consumption. If climate change is to be successfully curbed, this extraction must fall to near zero over the next 30–40 years. What does this mean for Saudi Arabia, whose national budget and prosperity depend entirely on oil revenues?

Well, Saudi Arabia not only has huge oil reserves, but also has two other resources in abundance: space and sunshine. Therefore, Saudi Arabia could play a very significant role in the future world market for synthetic fuels – just as it does in today's world market for fossil fuels.

If the country wants to produce the same 7 PWh of CO_2-neutral energy in 2050, it will need about 10–12 PWh of solar energy, for which a maximum of 50,000 km^2 of photovoltaic modules would be needed under the local conditions.[16] That is a good 2% of the country's surface area – more than half of which consists of uninhabited, desert-like areas anyway.

In other words, the foreseeable end of the fossil fuel era does not automatically mean the end of energy export-based prosperity in Saudi Arabia. Will the country seize this opportunity?

---

[16] 10-12 PWh correspond to a capacity of 5 TW PV modules. This is, of course, a rough calculation, intended only to show the dimensions and technical feasibility using PV power. It could be that, for example, CSP (Concentrated Solar Power) technology proves to be more suitable for the conditions in Saudi Arabia. However, CSP land requirements are of a similar order of magnitude to those of PV technology.

10
Raw Materials

Introduction

50 years ago, in 1972, a book was published which profoundly shaped the worldwide discussion on (ecological) sustainability and, in a certain sense, initiated this discussion in the first place: "The Limits to Growth" by Dennis Meadows and his colleagues.[1]

This work was groundbreaking in two respects: first, in *terms of content*, because for the first time systematic thought and research was conducted on the scarcity of raw materials, industrial production, environmental pollution, availability of food, population development and notably their interdependencies. Second in *terms of method*, because for the first time computer models were used to calculate the effects of these interdependencies and thereby to forecast future developments.

In "The Limits to Growth", particularly large space is devoted to mineral raw materials: the common metals such as iron, copper, aluminum, zinc, chromium; but also rare materials such as molybdenum or tungsten. In their so-called "standard world model," the authors predicted that (not

[1] At the time, Dennis Meadows was the head of a research group at the American Massachusetts Institute of Technology (MIT) that produced the findings on which the book is based.

© Springer-Verlag GmbH Germany, part of Springer Nature 2022

T. Unnerstall, *Factfulness Sustainability*,
https://doi.org/10.1007/978-3-662-65558-0_10

climate change, which was largely unknown at the time, but) a drastic decline in available raw materials in the course of the twenty-first century could lead to a global collapse of the world economy and, as a consequence, of the world population.

Ever since 1972, concerns about the scarcity of raw materials have been a central issue when it comes to the future of humanity. Examples are the discussion about "peak oil" (cf. Chap. 9) or the fear of phosphorus reserves drying up (with dramatic consequences for fertilizer and thus global food production). Recently, the question has been discussed whether there is enough lithium available to produce billions of batteries for electric cars in the future.

The fact is: Not only the consumption of fossil fuels, but also the global demand for raw materials has increased dramatically over the last 100 years. Despite the warnings of "The limits to growth", it has tripled again since 1970. And if in this century 10–11 billion people want to achieve a similar standard of living as in the West, the demand for iron, copper, aluminum, lithium, phosphorus, etc. will continue to rise.

So: When will humanity reach the limits of growth?

Basics

The earth is big, and its stocks of raw materials are almost unconceivably large. Even if only the earth's crust on the continents and here also only the upper 3 km – i.e. the part accessible to raw material mining today – is considered, this

Table 10.1 Raw material consumption to date. (Source: Wikipedia)

Raw material	Amount in the upper earth crust (in Gt)	Extraction to date (in Gt)	Extraction(in % of total amount)
Iron	56 million	40	0.00007
Aluminum	80 million	1.4	0.000002
Copper	60,000	0.8	0.001
Phosphorus	1 million	1.1	0.0001
Chromium	100,000	0.3	0.0003
Zinc	70,000	0.5	0.0007
Lithium	20,000	0.001	0.00005
Gold	4	0.0002	0.00005

material has a weight of over 1 billion Gt (Gt = 1 billion tons).[2]

The average abundance of the chemical elements and thus of the raw materials in this upper part of the Earth's crust is well known. It varies widely, ranging from $4 \times 10^{-7}\%$ for gold (corresponding to about 4 Gt of gold) to $2 \times 10^{-3}\%$ for lithium (= 20,000 Gt) to 8% for aluminum (= 80 million Gt) and silicon (27%).

If we compare these amounts with what mankind has extracted from the earth since the beginning of history, we get Table 10.1.

The result is striking: even in the case of copper and zinc, the "scarcer" raw materials, mankind has so far only used about 0.001% of the amount in the upper continental earth's crust.

In addition: Unlike fossil energy sources, mineral raw materials are generally not actually *consumed*, but only *used*, i.e. built into products and buildings. They are retained and can be recovered – in principle without any loss of

[2] Humans extract about 70 Gt from the earth every year, mainly 40 Gt of building materials (gravel, sand, etc.), 15 Gt of fossil fuels and 10 Gt of metal ores. Even if it continues to do so for the next 10,000 years, it will not even have moved 0.1% of this upper continental crust.

quality – at the end of life of these products/buildings. In other words, they can be recycled from waste.

This characteristic has two fundamental consequences. First, for most raw materials today, over 50% of the amount ever mined is still in use. Secondly, mankind is slowly moving towards a (largely) circular economy, at least for many raw materials: modern waste treatment technologies now achieve recycling rates of sometimes 50%, and for iron, aluminum, copper, etc. already 70–80%.

But then…where is the problem? How do these facts fit in with the forecasts and fears about impending raw material shortages?

Well, let us take **copper** as an example. Copper has an average abundance of only 0.006% in the earth's crust, but this is indeed just the average value: De facto, the concentration of copper in the earth's material varies very much – between 0% and 1%.[3] Of course, copper was mined in the past and is still mined today where the concentration is particularly high: because the lower the concentration of the raw material to be extracted in the earth's material, the more demanding, energy-intensive and therefore more expensive it is to actually extract it.

Against this background, the following question is important[4]: How much copper can be extracted based

- on the minimum concentration at which copper can still be extracted with *current* mining technologies, and
- on the *currently* known deposits that have this minimum concentration?

It is obvious that the answer to this question may change over time, as both mining technologies and geological knowledge of deposits evolve.

[3] In individual cases, there are also deposits with even higher concentrations.
[4] In technical jargon, it is also called the question of *copper resources*.

This leads to an interesting phenomenon: in 1972, when "The Limits to Growth" was written, the relevant analysis indicated that about 1.5 Gt of copper was still mineable. Since about 0.7 Gt of copper has been newly mined in the last 50 years, only 0.8 Gt of mineable resources should be left today, according to 1972-knowledge. At the current mining rate (20 million tons/year), these 0.8 Gt would be exhausted in 40 years, and the "Limits to growth"-forecast of drastically declining raw material resources in the twenty-first century would come true in the case of copper.

In fact, however, the forecast was wrong. Today, more than 5 Gt(!) of copper are expected to be minable in the future. Both the mining technology and the knowledge of deposits have improved considerably in recent decades. Moreover, with 60,000 Gt of copper actually present in the continental upper crust, it is quite clear that in another 50 years the situation will be similar.

Let's take a closer look now at the relevant data for the most important mineral raw materials in terms of volume: iron, aluminum, copper and phosphorus.

The Facts/Projections Until 2100

Iron

Iron is – along with wood and gravel – the most important raw material for mankind. For more than 2000 years, iron has formed the basis for almost all technologies: first in agriculture and for warfare, since the industrial revolution also in transport, households and for buildings.[5]

[5] Despite this long history: more steel is produced in one year today than in the entire history before 1900!

The Facts – Global View

- In the course of history so far, man has mined about 40 Gt (= 40 billion tons) of raw iron; more than three quarters of this in the last 50 years. Eighty percent of this material is still in use today, i.e. is in buildings, machines, cars etc. About 20% has been lost, i.e. is mostly scattered in landfills.
- Currently, annual production of raw iron (also termed pig iron) is about 1.3 Gt. Together with 0.5 Gt of recycling products (old scrap),[6] approx. 1.8 Gt of steel products are produced. The development over time is shown in Fig. 10.1.

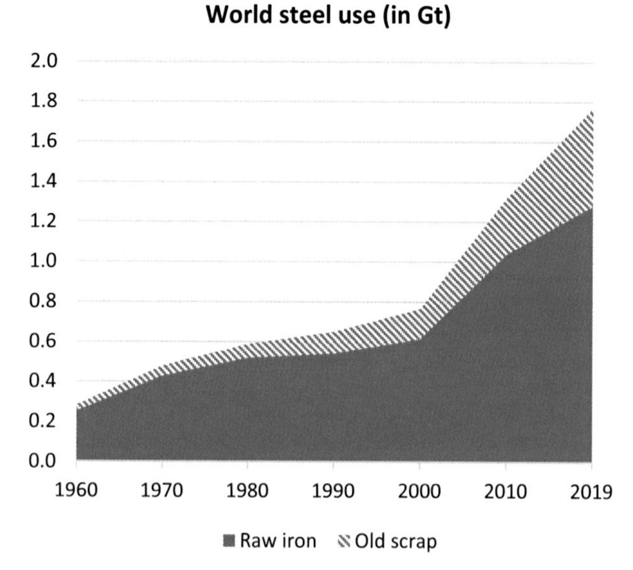

Fig. 10.1 World steel use in products and buildings, split between raw material (pig iron) and scrap, 1960–2019. (Sources: World Steel Association, USGS)

[6] In recycling, one distinguishes between two types of scrap. "New scrap" is metal waste already generated in the production process of iron or steel products; it is usually 100% recycled, i.e. returned to the production process. "Old scrap" is steel products at the end of their service life or steel in buildings that are demolished; it is only the recycling of this old scrap that ultimately matters when considering raw materials.

- Today, iron resources – i.e. iron that can be mined in known deposits using current technologies – are estimated at about 240 Gt. The actual amount of iron in the upper continental crust is more than 50 million Gt.
- The average life-time of steel-containing products is 40–50 years[7]; and the recycling rate in waste treatment of these products is currently in the order of 70–80%.

The Facts – The West

In the western countries, the use of steel[8] has been roughly constant for 50 years, with fluctuations, at 0.25–0.3 Gt/year.

However, the raw material required for this, i.e. pig iron, has been declining quite continuously and significantly since 1970 because increasing quantities of steel are being obtained from recycled material: Currently, it already accounts for 50% of total steel use (Fig. 10.2).

Therefore, although Western per capita steel use (in products/buildings) of ca. 300 kg/year remains well above the world level (200 kg), the annual average *raw material requirement* per capita of 150 kg is no higher than in the rest of the world.

Evaluation/Forecast

Over the past 50 years, annual steel use per capita in the West has fallen from 400 to 300 kg – despite significant

[7] In other words, current annual steel waste volumes are on the order of the annual steel use in the 1970s.

[8] It is important to mention that with "steel use", we denote here what is technically termed "true steel use" – meaning that the import/export-balances of steel-containing products are taken into account when calculating the steel use of a country or a region.

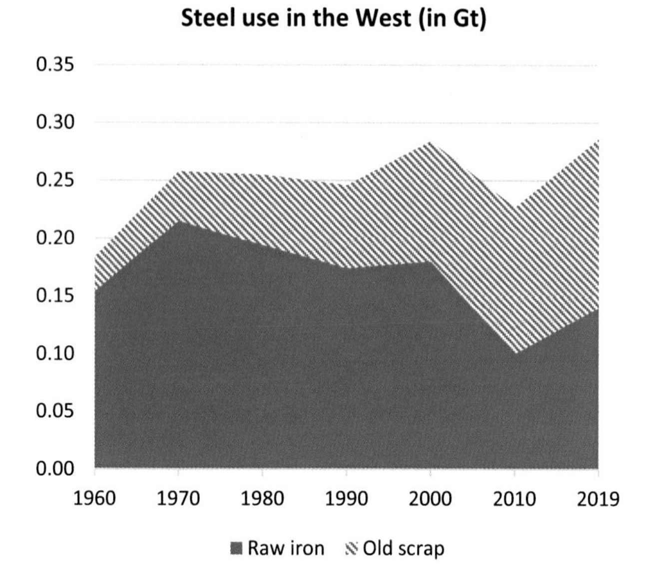

Fig. 10.2 EU + US (true) steel use in products and buildings, split into pig iron and scrap, 1960–2019. (Sources: World Steel Association, USGS)

increases in industrial production, standard of living, household appliance equipment, etc. Technological progress has thus significantly increased raw material efficiency in production processes; in some cases, steel has also been replaced by other materials. With regard to the future foreseeable today, it can be assumed that, (a) this trend will continue and, as a result, steel use in the West will remain at 250–300 kg per capita; and, (b) 10–11 billion people will achieve a similar standard of living in the twenty-first century. Consequently, world steel use can be expected to increase from about 1.7 Gt today to about 3 Gt in the next decades.

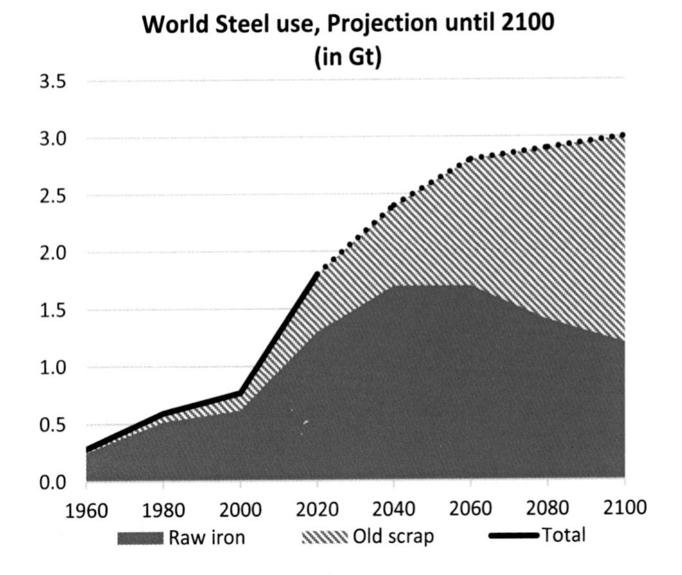

Fig. 10.3 World steel use in products/buildings, split into pig iron and scrap, from 2020 forecast values. (Sources: World Steel Association, Fraunhofer (2016), own calculations)

At the same time, it is possible to estimate how the recycling quantities will develop.[9] Based on these assumptions, Fig. 10.3 shows that the demand for iron raw material is likely to reach a peak around the middle of the century and then slowly decline again. In other words: with respect to iron, mankind is indeed moving toward a circular economy.

This projection entails that another 100–150 Gt of iron will be mined in the twenty-first century.

What do these figures mean?

Even if we take *today's* knowledge of iron deposits and *today's* extraction technologies as a basis, by 2100 mankind will have extracted just about half of the iron resources; and the majority of these quantities of iron will continue to

[9] The German Fraunhofer Institute UMSICHT published a comprehensive study on this several years ago: Fraunhofer (2016).

circulate, i.e. continue to be usable. Quite apart from the enormous iron amounts stored in the upper earth's crust, there is absolutely no question of a shortage of iron as a raw material.

Copper

The use of copper goes back even further than that of iron. The Bronze Age began 4000 years ago (bronze consists of 90% copper), when for the first time a metal became the focus of mankind's cultural and technological development. Today, copper is a very widely used raw material – especially in cars, machinery and energy technology.

The Facts – Global View

- To date, about 750 million tons of copper have been mined, of which about 500 million tons are still in use.
- Currently, the annual production of new copper is about 21 million tons; together with about 4 million tons of recycled copper (from waste products), about 25 million tons of copper products leave the factories. The development over time is shown in Fig. 10.4.
- Current copper resources – according to current extraction technologies and known or suspected deposits – amount to about 5600 million tons. Total copper in the upper continental crust amounts to 60,000,000 million tons.
- The average lifetime of copper-bearing products is 30–40 years. Currently, about 30–40% of the copper waste stream is recovered and returned to the production process.

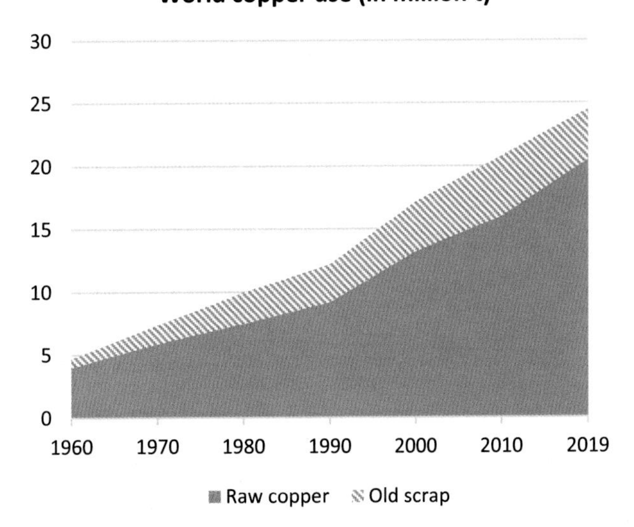

Fig. 10.4 Use of copper in the world, divided into raw material and scrap, 1960–2019. (Sources: World Copper Factbook, USGS, Glöser et al. (2013), International Copper Association)

The Facts – The West

In Western countries, the use of copper has been declining for about 20 years and is now at 5–6 million tons/year. The raw copper required for this purpose has also been declining for several decades because copper recycled from waste products is increasingly being used. Currently, this ratio is about 50% (Fig. 10.5).

The per capita use of copper in the West is significantly higher than the world average; however, the demand for the raw material copper per capita and year is currently about 3.4 kg, which is only 25% higher than in the rest of the world.

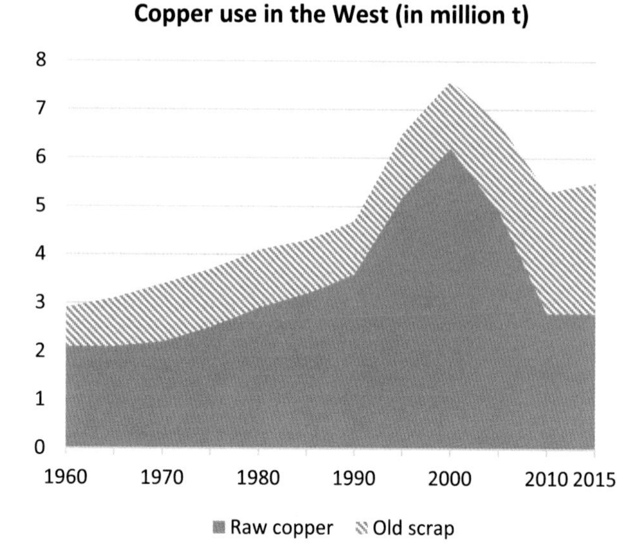

Fig. 10.5 Use of copper in EU + USA, split between raw material and scrap, 1960–2015. (Sources: World Copper Factbook, USGS, Soulier et al. (2014), Copper Development Association (2017), own calculations; figures before 1990 partly estimated/interpolated)

Evaluation/Forecast

Global copper use is very likely to rise sharply. Assuming a Western level of consumption for 10–11 billion people, we are looking at 60–70 million tons/year. At the same time, the trend of increasing recycling rates already observed in recent decades will continue. On this basis, it is possible to make a forecast – similar to that for iron – for the expected copper raw material demand, i.e. for copper mining in the twenty-first century.

Copper mining is expected to reach a peak in the order of 40 million tons in the second half of the twenty-first century and then slowly decline again due to further increasing recycling quantities: Fig. 10.6.

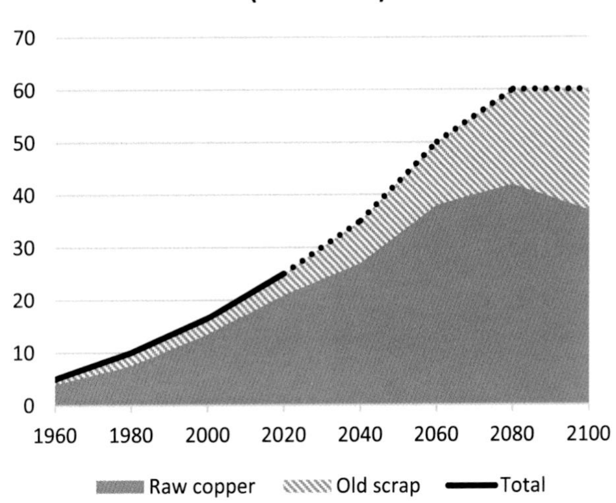

Fig. 10.6 Use of copper in the world, 1960–2100, divided into raw material and old scrap; forecast values from 2020. (Sources: World Copper Factbook, USGS, Glöser et al. (2013), International Copper Association, own calculations)

In total, another 3000 million tons of copper could be mined until 2100 – about half of the resources according to current knowledge, and only a tiny fraction of the total copper available.

Aluminum

At 8%, aluminum is the third most abundant element in the earth's crust after oxygen and silicon. It was discovered only about 200 years ago and has been used on an industrial scale for about 100 years. In recent decades, however, the use of aluminum – being significantly lighter than iron but equally sturdy – has virtually exploded: in buildings, cars, machinery and also in packaging materials.

The Facts – Global View

To date, man has mined about 1500 million tons of aluminum, more than three quarters of this in the last 30 years.

- About 1100 million tons of this aluminum is still in use, the rest (400 million tons) has been lost. The development over time is shown in Fig. 10.7.
- Currently, annual extraction is about 65 million tons; together with about 20 million tons of recycled material (old scrap), this is used to produce about 85 million tons of aluminum products (and, to a small degree, for other uses).
- Today, aluminum resources are estimated – based on current extraction technologies and currently known deposits – at 12,000 million tons; however, many millions of Gt are stored in the upper earth's crust.

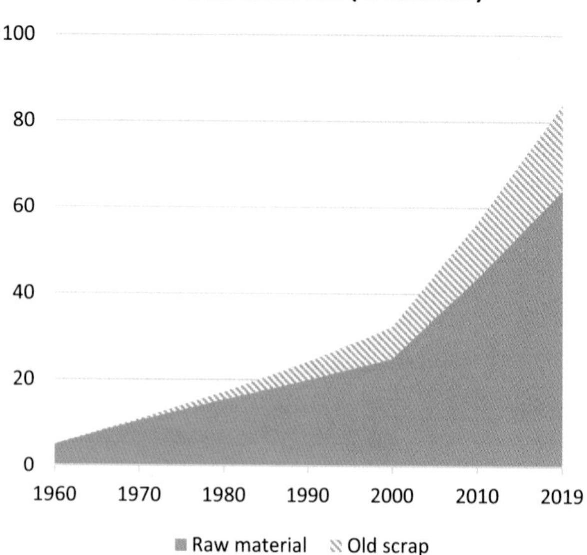

Fig. 10.7 Aluminum use in the world, split between raw material and old scrap, 1960–2019. (Source: IAI)

- The average life-time for aluminum products is in the order of 30 years,[10] and the current recycling rate from aluminum-containing waste streams is around 75% on world average.

The Facts – The West

Unlike in the cases of iron and copper, there is no sign yet of stagnation in the use of aluminum in Western countries[11]. In recent years, the use of aluminum has once again increased significantly, as Fig. 10.8 shows. The *demand for raw material,* however, has been increasing only slowly over the past 15 years due to high recycling rates, and will therefore very likely decline in the coming decades.

With these values, annual per capita use of aluminum is about twice the world average (22 kg vs. 11 kg), and the raw material demand is also significantly higher.

Evaluation/Forecast

Global use of aluminum will continue to increase sharply in the twenty-first century; based on today's per capita level in the West for 10–11 billion people, it is likely to exceed 200 million tons/year in the second half of the twenty-first century. At the same time, however, recycling rates worldwide will rise to around 80% by 2050, according to a forecast by the World Aluminium Association. On this basis, the

[10] The current waste volume of just under 30 million tons is therefore roughly equivalent to the use of aluminum (in products) in the 1990s.

[11] Due to the availability of data, we do not use the EU + USA as a basis for the Western countries in this case, but Europe + North America. However, the EU + USA are responsible for the vast majority of the figures shown here.

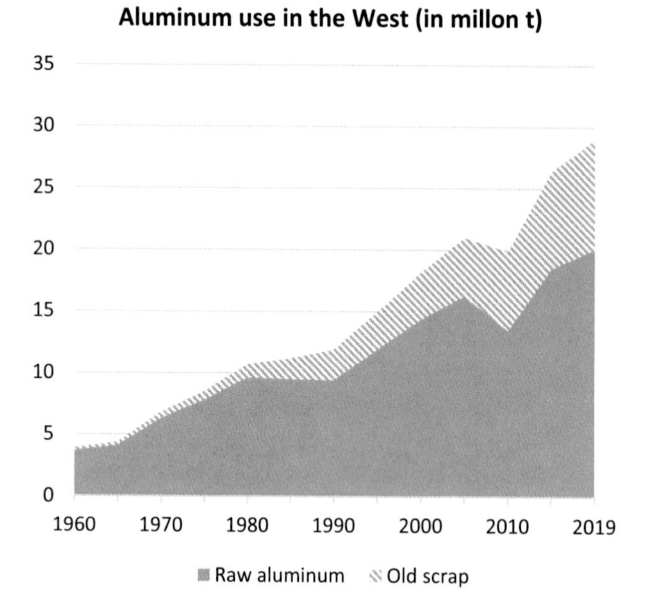

Fig. 10.8 Aluminum use in products in North America and Europe, split between raw material and scrap, 1960–2019. (Source: IAI)

development of raw aluminum demand up to 2100 can be roughly estimated, as illustrated in Fig. 10.9.

Hence, aluminum in the order of 8000 million tons will be mined over the next 80 years. This means that even on the basis of known deposits that can be extracted with today's technologies, there will be no shortage of aluminum in the twenty-first century. The total amount of aluminum in the upper earth's crust is inexhaustible anyway.

Phosphorus

Phosphorus is a fundamental component of all life, and in particular it is indispensible for the growth of plants. In the form of artificial fertilizer, phosphorus has played a crucial

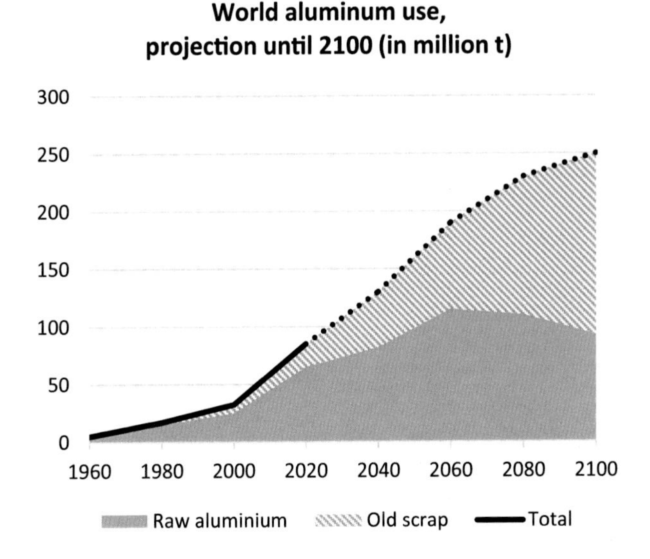

Fig. 10.9 Aluminum use in products in the world, 1960–2100, divided into raw material and scrap, from 2020 forecast values. (Sources: IAI, own calculations)

role since about 1950 (along with nitrogen) in multiplying agricultural yields worldwide and thus feeding the rapidly growing world population. From today's perspective, there is no alternative to phosphorus in this context – without phosphorus fertilizers, it is impossible to adequately supply the soon to be 10–11 billion people with food. Therefore, the availability of this raw material is of utmost importance.

A few years ago, the question of how long phosphorus reserves will last and whether "peak phosphorus" is to be expected in the next few decades – i.e. the point in time at which world phosphorus production would irrevocably decline – was seriously discussed in this context.[12]

[12] E.g., in Bardi (2013). Since huge, easily exploitable phosphorus deposits were newly discovered in Morocco in 2011, this discussion has largely fallen silent.

The Facts – Global View

- Over the last 100 years, mankind has mined about 1100 million tons of phosphorus;[13] about three quarters of this has been used for fertilizer, i.e. applied to arable land.
- Via this pathway, a considerable part (400–500 million tons) has entered the earth's water system and thus ultimately mainly the oceans (cf. Chap. 15). The rest is dispersed in the environment.
- At present, a little over 30 million tons of phosphorus are extracted per year, of which about 20 million tons go into fertilizer production. The development over time is shown in Fig. 10.10. After several decades of stagnation, phosphorus production has increased significantly again

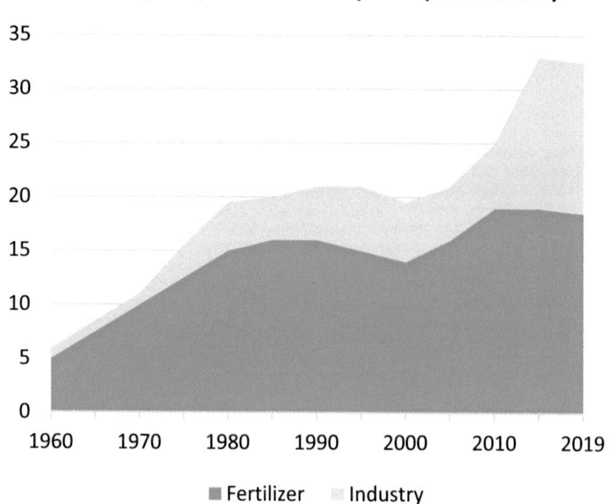

Fig. 10.10 Phosphorus consumption in the world, 1960–2019, in million tons. (Sources: OurWorldinData, Chen and Graedel (2016))

[13] Another 1300 million tons or so of phosphorus were extracted from the deposits in this process, but are found in the residues of mine production.

in recent years, also for non-agricultural applications (e.g. batteries).

- Recycling of phosphorus is possible in principle, but is still in its infancy; efforts to this end are only slowly developing.
- Current resources are estimated at over 30,000 million tons (30 Gt). A total of about 1 million Gt is stored in the upper continental crust.

The Facts – The West

The use of phosphorus in fertilizers has been declining in the West for decades, and it stands today at around 3 million tons/year or 3.5 kg/capita (Fig. 10.11). This figure is expected to continue to decline slightly in the future as agricultural practices continue to improve and phosphorus recycling slowly increases.

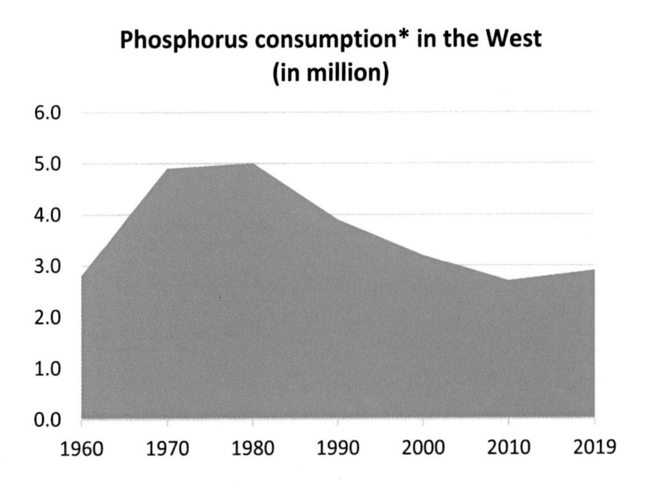

Fig. 10.11 Phosphorus consumption in North America + Europe, 1960–2019, in million tons. (Sources: OurWorldinData, Chen and Graedel 2016) *for fertilizer only

Evaluation/Projection

Global use of phosphorus for fertilizer could increase to 35–40 million tons/year (from 20 million tons today) during the twenty-first century, equivalent to current Western per capita consumption times 10–11 billion people. There is some evidence to suggest that it will remain lower – mainly the fact that, according to the findings of recent years, many agricultural soils are oversupplied with phosphorus. Significantly more phosphorus is applied to arable land than is taken up by the plants. In the best case, therefore, up to 10 million tons/year could be saved with improved agricultural methods.

But even a peak annual consumption of about 40 million tons would only lead to a further resource depletion of max. 4000 million tons by the year 2100 – with current resources of 30,000 million tons, this does not cause any problems. The future of other industrial uses of phosphorus, and hence the total future demand for this raw material, is completely uncertain; but even if present quantities are multiplied for new uses, there will be no shortage of phosphorus in the twenty-first century.

Evaluation/Summary

A major argument for the often repeated paradigm that "Western consumption and its growth-oriented economic system are unsustainable" is raw material consumption. If every inhabitant of the earth had a similar consumption of iron, copper, aluminum, etc. as the EU or US citizen, the argument goes, the supplies would soon be exhausted: humanity would quickly reach the limits of growth.

The data in this chapter have shown that this reasoning is wrong. It does not take into account three essential aspects:

- First, economic growth does not automatically entail increasing use of mineral raw materials. Rather, at some point in economic development and standard of living – clearly evident from the relevant data for the West – there are saturation tendencies. Further economic growth is then driven primarily by the service sector, and many innovations go in the direction of higher raw material efficiency.
- Second, recycling. The developed industrial nations now achieve considerable recycling rates – certainly less high than would ideally be possible, but still high enough so that, in many cases, raw material demand has actually been falling again for decades due to the increasing use of recycled material.
- Finally, there is the mistake of equating "raw material stocks" with "reserves that can be mined with current technologies in currently known deposits," without taking into account the ongoing progress in this respect. In other words, the figures published as "raw material resources" always represent only a snapshot that reflects the current state of knowledge.

Therefore – at least for the raw materials considered here in more detail – the opposite is true. Even if every inhabitant of the earth has (and possibly will have in the course of the twenty-first century) the same per capita raw material demand as an inhabitant of the West, in a world of 10–11 billion people: there will be no shortage of raw materials in the twenty-first century.[14]

[14] Real scarcity is meant here; temporary shortages on world markets may occur, cf. below.

Often, the argument has been and is still being put a different way: sufficient raw material stocks may still be available in the future, but they would become increasingly expensive in view of declining raw material concentrations in the deposits, and further extraction would therefore no longer be affordable. These forecasts have not come true either, and in all probability will not come true in the future. In economic terms, the situation is as follows:

It can and it does happen that the stocks of a certain raw material that can be mined with *current technologies* in the *currently known deposits* are foreseeably running out. This then regularly leads to rising world market prices (demand foreseeably continues to rise, supply foreseeably falls). In such cases, the following alternatives exist for the respective markets:

- New deposits are sought at greater expense than in the past, and/or new technologies are developed at greater expense than in the past, in order to be able to use new deposits with possibly lower raw material concentrations.[15]
- Recycling efforts are intensified to extract more material from waste streams.
- Substitutes are employed
- Ways are developed to use less raw material for the same products, i.e., to increase raw material efficiency.

[15] There is the objection here that, as raw material concentrations decrease, these technologies will increasingly have to work with chemical substances and will therefore tend to be increasingly harmful to the environment. This will be true in some cases, but only underscores the need – which exists anyway – for high safety and environmental standards to be mandated in raw material mining. The current environmental impacts of commodity mining are well documented (World Atlas of Desertification (EU 2018), Chapter "Mining"): today, only 0.05–0.1 million km^2, a relatively small area, is affected.

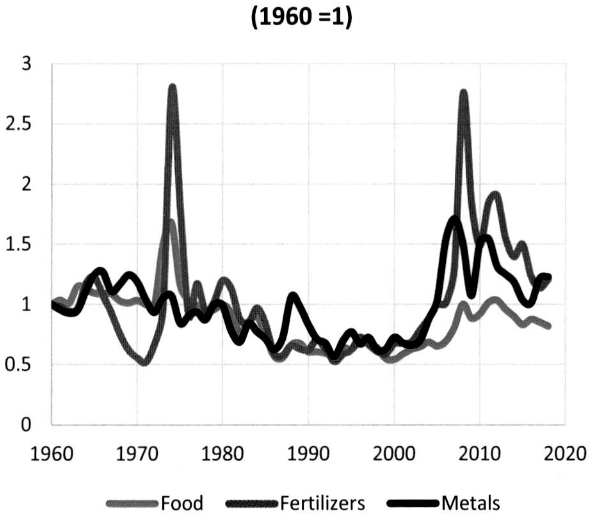

Fig. 10.12 World market prices for food, fertilizers, metals 1960–2018, inflation-adjusted, 1960 = 1. (Source: World Bank)

Regularly, one or more of these options lead to an increase of supply and/or a decrease of demand, and thus world market prices fall again. Indeed, world commodity prices – after considerable upward and downward fluctuations – are today almost exactly at the level of 1970 again (adjusted for inflation; see Fig. 10.12).

A final variant of the argument is[16] this. The future extraction of raw materials is foreseeably limited by the fact that the energy required for extraction would simply be too high – in view of the abundance of regenerative energies available, this is not a convincing argument either.

The point remains: there is no serious scarcity of important raw materials today, nor in 2100, nor – as far as one can rationally assess at all – in the next centuries.

[16] For example, in Bardi (2013).

Man has not plundered the planet of its raw materials, he has just scratched the surface a little bit. The generations of the twentieth century have certainly made it somewhat more difficult for future generations to extract the raw materials (because many of the best deposits have already been exploited); but in return they have also passed on to these generations much more advanced technical possibilities and a much more detailed geological knowledge than they themselves had at their disposal.

To put it another way: there are obviously „limits to growth" on a finite planet as far as human exploitation of mineral resources is concerned. But these limits are far, very far away.

Excursion: The Alleged Lithium Shortage

Time and again, right up to today, there have been headlines claiming that the triumphant advance of battery-powered electric cars could be slowed down by shortages of raw materials. In particular, it is said, there could be shortages of lithium as an essential component of lithium-ion batteries (currently the leading technology in batteries for e-cars).

A brief consideration involving the relevant figures shows that these fears are unfounded. Let us assume for a moment,

- lithium-ion batteries will remain the most important battery technology in the coming decades;
- the lithium content in the batteries – max. about 8 kg per battery – cannot be reduced (despite current research to this end);
- the recycling rate of lithium in batteries cannot be pushed above 50%.

(all three rather unlikely assumptions). Under these assumptions, future e-car batteries contain about 4 kg of new lithium and 4 kg of recycled lithium. We assume 2.5 billion cars in 2050 (today there are about 1.2 billion) and consider a scenario with 100% e-cars. Assuming a lifetime of 10 years for batteries, we would then need a production volume of 2.5 billion × 4 kg × 0.1 = 1 million tons of lithium every year.[17] Although this is a lot compared to today's mining volume, it is not a fundamental problem: Even the currently estimated resources of lithium (known deposits, mineable with today's technologies) are more than 60 million tons, i.e. would last until about the end of the century under the above assumptions.

The total lithium reserves in the upper earth's crust amount to 20,000 billion tons (!); and the deposits in seawater (250 billion tons) are also de facto unlimited.

[17] With more realistic assumptions – halving of the lithium content, recycling rate at 75% – it is only 0.25 million tons/year. This means that even the currently known lithium resources will last for more than 200 years.

Part IV

Ecological Hotspots

11

The "Ecological Footprint"

Introduction

There are only a few concepts that originated in the natural sciences that have been adopted into common usage. The concept "ecological footprint" is one of them. It was invented 25 years ago and has since become probably the most popular "*sustainability indicator*": It is regularly used in the media and in many political discussions – especially around the annual "Earth Overshoot Day" mentioned in the introduction – to demonstrate that humanity is (allegedly) overusing the earth's resources. What is more, the fact that the ecological footprint of the USA and the EU is significantly higher yet than the world average is one of the most important arguments in the increasing fundamental criticism of the Western lifestyle and economic system.

The success of the concept "ecological footprint" is undoubtedly due to the fact that it combines very different aspects in the human-nature relationship into a single figure and therefore allows short, striking statements. However, this success has come at a high price: the *concept* has been adopted from science into societal discourse, but along the way the *scientific definition* and thus the actual meaning of this indicator has been lost. The consequences are serious. The scientific results regarding the ecological footprint are

© Springer-Verlag GmbH Germany, part of Springer Nature 2022

T. Unnerstall, *Factfulness Sustainability*,
https://doi.org/10.1007/978-3-662-65558-0_11

commonly misinterpreted, and arguments and accusations based on them are therefore usually unfounded.

In this chapter, I will first explain the correct meaning of the concept. On this basis, I then answer the question of whether or not the ecological footprint of humanity or the West is really a problem, using the most important scientific data in this regard.

So what does the term "ecological footprint" mean?[1]

At its core, this concept is about the ratio of resources *available on Earth* to those *claimed by humans*. Now, planet Earth possesses a whole range of resources that are important for humans: air, water, soil, solar energy, biomass (plants and animals), fossil materials (coal, oil, natural gas), mineral raw materials, etc. In the conception of the ecological footprint, the scope is confined in a *first step* to the biomass resources. In a *second step*, the consideration is focused on the "sustainable" part of these biomass resources, i.e., on those quantities that grow back or are newly formed each year.

In a *third, decisive step*, the scope is even further restricted: It is not about all (annually renewable) biomass resources of the earth, but only about those that are actually available for humans *within the framework of their current economic practices*. For example, the resource "fish" is not about all fish in the world's oceans, but a) only about the fish species actually fished by humans (i.e. 1600 of 35,000 fish species), and b) only about inland and coastal waters; because 95% of fishing takes place in these waters.[2] The resource "wood" does not refer to the entire vegetation in the forests, but only to the stem wood, because only this is used to a greater extent by humans.

[1] The following discussion is based on the research literature on the ecological footprint, especially Wackernagel et al. (2005). The data in this chapter are taken almost exclusively from the Global Footprint Network database.

[2] These waters account for less than 10% of all waters on Earth in terms of area.

The global biomass resources defined in this way are referred to by the term "**biocapacity**": the biocapacity of the earth in a given year is thus the amounts of those types of biomass newly formed that year on earth that are important for (current) human economic activity. Biocapacity is divided into four categories[3]:

- Arable products (official term: "cropland" = biomass grown on cropland)
- Pasture grass (official term: "grazing land" = grass on agricultural pastures)
- Wood (official term: "forest products" = stem wood in forests)
- Fish (official term: "fishing grounds" = fished species in inland/coastal waters)

These four categories are converted into a single measure by a rather complex methodology (which we will not go into in this book): the GHA ("Global Hectars").

Biocapacity is thus an *anthropocentric* (i.e., tailored to human activities) concept that actually encompasses only a fairly small fraction of the Earth's total renewable biomass. Therefore, it is not a fixed, temporally constant quantity; on the contrary, the biocapacity depends on the development of human technology and human economic habits. This is particularly true for biocapacity in the category of "arable products": it has increased sharply over the past 60 years due to more cropland and increased agricultural yields (resulting from fertilizer use etc.). Correspondingly, new cropland leads to more biocapacity, while deforestation leads to less. In this way, the overall biocapacity – as defined by the

[3] There is a fifth category, "building land"; it refers to the biomass that could theoretically be produced by agriculture on land that is in fact used for buildings and transport. Because of its quantitative insignificance, I will consider it here only in passing.

"ecological footprint" concept – is now about 30% higher than it was in 1960.

Thus, while *biocapacity* refers to the biomass resources sustainably usable in the framework of current human agriculture, forestry, and fishery, the "**ecological footprint**" is defined as the amount of biomass resources that humanity actually uses in a given year. In detail:

- Arable products:

  In this category, the ecological footprint is equal to the biocapacity because humans generally harvest all fields and also consume the agricultural products harvested.

- Pasture grass:

  Here, the ecological footprint captures how much of the grass annually (re-)growing on global pasture lands is actually eaten by livestock. This amount may or may not exceed biocapacity.

- Wood:

  The ecological footprint measures the annual (stem) wood use. It may be higher than the corresponding biocapacity: in this case, more wood is harvested from forests than regrows in the year considered.

- Fish:

  The ecological footprint in this category measures the global annual fish catch. Again, this amount may exceed biocapacity: In this case, fish stocks in inland and coastal waters included in the analysis are declining.

Unlike in biocapacity, however, for the total ecological footprint a further category is added, called the "*carbon footprint.*" It is defined as the forest biomass that would be required to fully absorb the annual amount of CO_2 emitted by humans into the atmosphere via photosynthesis, thus avoiding (the accumulation of CO_2 in the atmosphere and thus) the greenhouse effect.[4] Hence, unlike the other categories, this is not a *real* amount of biomass actually consumed, but rather a *fictitious* amount – precisely the biomass that would be mathematically necessary for human CO_2 emissions to have no effect on the climate. This conversion of CO_2 emissions into biomass (and ultimately into the measure GHA) makes it possible to express the problem of man-made climate change as a burden on natural biomass resources and to integrate it into the concept of the ecological footprint.

This methodology has a significant consequence that is often overlooked. If we now calculate the individual biomass consumptions in each of the total five categories, we find that the carbon footprint dominates the other four categories: it currently accounts for about 60% of humanity's total ecological footprint. This obscures the important question of whether or not the *real* biomass use – that is, use of arable products, pasture grass, wood, and fish – exceeds the sustainably useable resources (see Introduction).

Therefore, in order to assess the real biomass consumptions, I exclude climate change in this chapter. In other

[4] For a more detailed explanation: There are three paths for the annual CO_2 emissions of mankind:

- A part is absorbed by the oceans (photosynthesis of plankton).
- A part is absorbed by the plants on the land surface of the planet.
- The rest remains in the atmosphere, leads to an increased CO_2 concentration there and thus contributes decisively to the greenhouse effect, i.e. to climate change.

On this basis, one can calculate how much forest (as the most important form of land plants in terms of quantity) would be required to bind *all the* CO_2 (not absorbed by the oceans) instead of the part mentioned above. This forest area is defined as the "ecological footprint carbon."

words: I consider only the *ecological footprint without the category of carbon*[5] and compare it with the biocapacity.

So far, we have understood the terms "biocapacity" and "ecological footprint" to refer to the entire earth and humanity as a whole, respectively. However, they can just as well be applied to world regions and to individual countries. In this case, of course, when calculating the ecological footprint in the four categories (arable products, pasture grass, wood and fish), one must take into account the respective imports and exports.[6]

Summarizing, the central issue in this chapter is this: Are natural biomass resources being overused or not – i.e. is the (non-energy) ecological footprint larger or smaller than the biocapacity? We will answer this question for the world as well as for the West.

The Facts – Global View

The current ecological footprint – biocapacity ratio in the four categories is depicted in Table 11.1.

Mankind thus uses about 70% of the available biomass resources that are newly formed in the relevant economic areas. The temporal development of the ratios in the individual categories is shown in Fig. 11.1. While the utilization rate of wood resources and pasture grass has been largely constant for about 30 years (because wood production and livestock have grown only slightly since 1990),

[5] In his book "2052" (Randers 2012), Jorgen Randers coined the term "non-energy ecological footprint" for this, which is quite apt.

[6] In actual implementation, these calculations are very complex because many factors have to be taken into account when converting biomass quantities to the one measure GHA: the different yields per hectare in the regions/countries, the different wood growth rates, and many others.

Table 11.1 Utilisation of Biocapacity, world, 2018. (Source: Global Footprint Network)

Category	Ecological footprint/ biocapacity (%)
Arable products	100 (by definition)
Pasture grass	70
Wood	42
Fish	64
Total	**69**

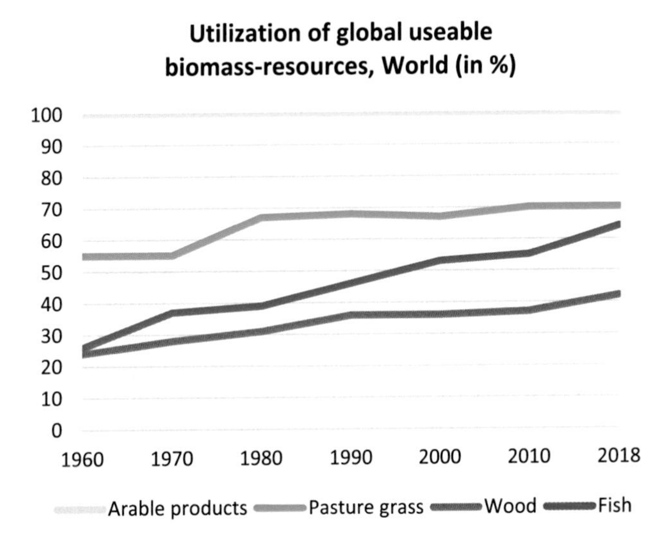

Fig. 11.1 Ecological footprint in relation to usable biocapacity in four categories, world, 1960–2018, in %. (Source: Global Footprint Network)

fishing (including aqua production) has been steadily increasing.

The overall picture of the ecological footprint (excluding carbon) and of the global biocapacity is shown in Fig. 11.2. Humanity's biomass resource consumption has roughly doubled in the last 60 years; and the utilization rate, i.e., the ratio to usable biocapacity, has increased from about 40% to almost 70%.

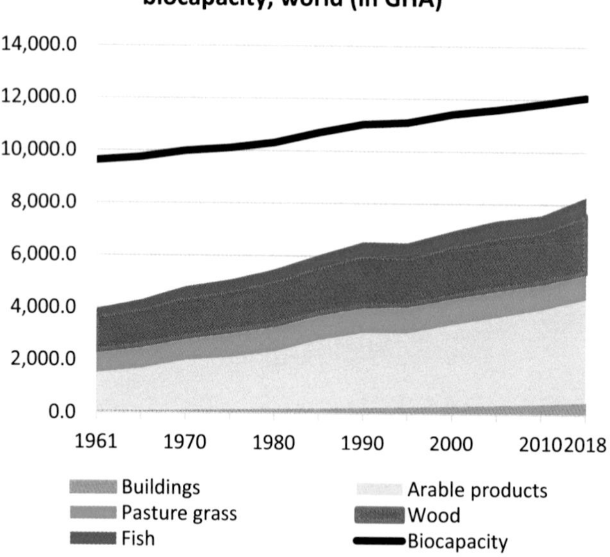

Fig. 11.2 Total ecological footprint (excluding CO_2) and biocapacity of the world, 1960–2018, in GHA. (Source: Global Footprint Network)

The Facts – The West

The overall picture for the West – Europe (excluding Eastern Europe) and the USA taken together[7] – is not very different. The ecological footprint as a whole is well below the domestic biocapacity, as Table 11.2 shows:

[7] Figures for the EU are not readily available from the Global Footprint Network database; therefore, I use the figures for Europe (excluding Eastern Europe) here.

Table 11.2 Utilization of biocapacity in the West, 2018. (Source: Global Footprint Network)

Category	Ecological footprint/biocapacity (%)
Arable products	97
Pasture grass	162
Wood	64
Fish	50
Total	**81**

Here, among other things, the assessment is confirmed that we already know from chap. 4 (Food): The West is a net exporter of arable products.[8]

A second important fact can be derived from Fig. 11.3: While in the world as a whole the ecological footprint is increasing, it has stabilized in the West for the past 20 years. The reason for this is easy to understand: Food consumption – grain, meat, fish – has only slightly gone up since 1990, while wood consumption has declined significantly in recent decades.

Projection Until 2050

What will the further development of the global (non-energy) ecological footprint look like, with the world population increasing to 10–11 billion people?

This question is easier to answer than it might seem:

[8] The reason for the approx. 160% in the use of grazing land – i.e. in the consumption of products derived from livestock (beef, sheep meat, dairy products, wool) – is unfortunately not understandable on the basis of the published data. 160% actually means that significant quantities of these products are imported. However, the West has a balanced export/import balance of beef and sheep meat, and it exports (net) significant amounts of dairy products (cf. footnote 6, Chap. 7)

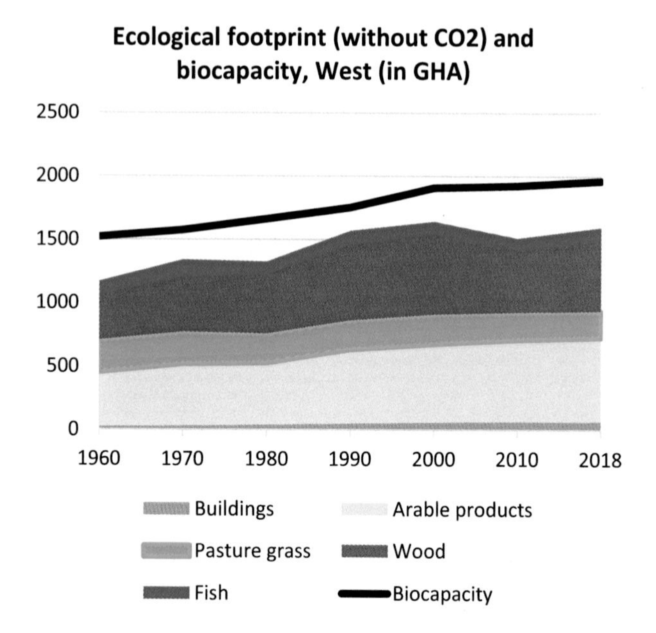

Fig. 11.3 Total ecological footprint (excluding CO_2) and biocapacity of Europe (excl. Eastern Europe) + USA, 1960–2018, in GHA. (Source: Global Footprint Network)

- Arable products consumption will continue to increase: on the one hand, through a moderate expansion of cropland, and on the other, through an increase in yields per hectare, especially in Africa and parts of Asia (cf. Chap. 7). By definition, biocapacity will increase by the same amount.
- Pasture grass consumption has been stagnant for 30 years and for this reason will increase only moderately, if at all. Biocapacity here is constant (as it has been since 1960).
- Wood use will tend to decrease. Currently, one third of the world's wood demand is still for heating and cooking (especially in Africa and Asia); and it can be assumed that this will be successively replaced by more modern

Table 11.3 Global biocapacity utilization, projection 2050/2070. (Sources: Global Footprint Network, own calculations)

Category	2018 (%)	2050/70 (projection) (%)
Arable products	100	100
Pasture grass	70	70–75
Wood	42	45–50
Fish	64	85–90
Total	**68**	**75–80**

forms of energy in the course of further economic development. However, we assume constant use here, together with a slightly decreasing biocapacity due to ongoing forest loss (cf. Chap. 13).

- Fish consumption will certainly continue to increase (but almost exclusively by aquaculture). We assume here that it will grow again by 2050 at a similar rate as in the last 30 years. Biocapacity should remain constant here (as it has since 1960).

If we summarize these developments,[9] we obtain Table 11.3.

In other words, even in the second half of the twenty-first century, humanity will most likely not exhaust the usable biocapacity – despite the expected growth of the world population to 10–11 billion people (see also Fig. 11.4).

[9] The attentive reader may wonder why I do not use the per capita values of the West for the future forecast here – as I did for energy and raw materials. Indeed, it is possible to calculate per capita ecological footprints for the world and, of course, for regions and countries. The problem, however, is that these values are not comparable with each other (unlike, for example, steel consumption per capita) because of the complex methodology used. For example: food consumption per capita in the West is about 20% higher than the global average, but the ecological footprint per capita in the „arable products"category is 70% higher than the global value; accordingly, the biocapacity "arable products" in the West is also valued higher in relative terms. For this reason, I chose this path for the forecast.

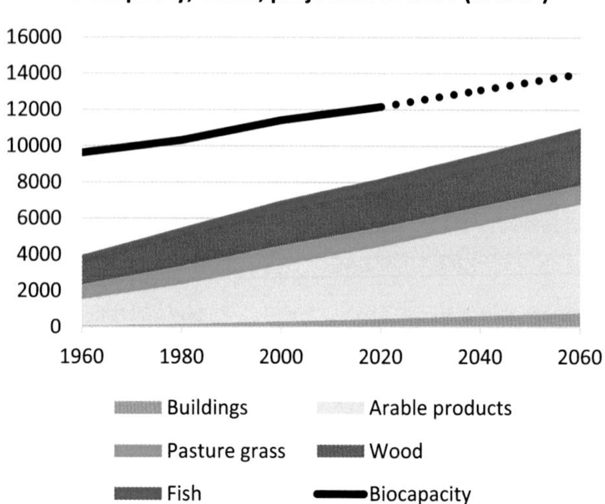

Fig. 11.4 Ecological footprint (excluding CO_2) and biocapacity of the world, 1960–2060, from 2020 forecast values, in GHA. (Sources: Global Footprint Network, own calculations)

Evaluation/Summary

You probably feel much the same way I do. With the term "ecological footprint of mankind" you associate the idea that it means *all* impacts caused by humans on the ecosystems of the planet: depletion of raw material stocks; environmental consequences of agriculture and forestry; decimation of fish stocks; pollutants in air, water and soil; extinction of animal species; garbage on land and in the oceans; deforestation; ozone hole and of course CO_2 emissions/climate change, just to name the most important aspects.

However, this idea does not hold true. The "ecological footprint," as it is scientifically defined and thus forms the basis for all media reports on this topic, captures only a small part of this long list. It essentially includes two items: biomass consumption (i.e., the direct impacts of agriculture, forestry and fishery) and CO_2 emissions (i.e., primarily the current impacts of energy use). Accordingly, two key insights can be derived from the scientific results obtained in the context of the ecological footprint concept:

- The message regarding CO_2 emissions/climate change is that the earth's forests are nowhere near sufficient to absorb man's CO_2 emissions. Therefore, a considerable part of this CO_2 remains in the atmosphere every year and thus accumulates (and the consequently successively increasing CO_2 concentration in the atmosphere leads to climate change). However, this insight is really not new – it was clear to climate scientists well before the concept of the ecological footprint was created.
- The message regarding biomass consumption is: this consumption has doubled since 1960, but it is still significantly below the annual growth of arable products, pasture grass, wood and fish. In other words: In terms of this concept, human agriculture, forestry, and fishery are *sustainable*[10] – they use fewer (renewable) resources than are available. Moreover, it can be assumed with a high degree of certainty that this will continue to be the case in the future, with a world population of 10–11 billion people. The earth provides enough resources to feed such a large humanity in a good and healthy way. (This was already the summary in Chap. 4).

[10] Other ecological effects of agriculture – especially soil degradation, nutrient surpluses with consequences for the oceans, pesticides with consequences for wildlife, etc. – could modify this judgment, but are not captured by the concept of "ecological footprint."

Regarding the West, it has become a commonplace that it claims resources in an unsustainable way – but this (pre-) judgment is simply refuted by the data (in the respect discussed here). It is a major contributor to climate change, yes, but with their real resource consumption in terms of food, wood and fish, Europe and the USA live within the framework of sustainability.

Excursion: China's Ecological Footprint

The world as a whole and the West in particular are living sustainably – in terms of the concept "non-energy ecological footprint", i.e., in terms of their use of agricultural products, fish and wood. This is the central message of this chapter. The same is true for most individual countries and regions, but not all. The countries with the largest true "overshoot" – that is, the largest biomass consumption not covered by their own biocapacity – are China and India.

China, in particular, consumes more arable products, meat from livestock and fish than its domestic resources can provide: the country lives off imports from other regions to about 20%, as shown in Table 11.4.

This deficit has emerged in the last 20 years. It is also not problematic in the framework of world trade and globalization, especially since China has large export surpluses in

Table 11.4 Ecological footprint/biocapacity ratio in China, 2018. (Source: Global Footprint Network)

Category	Ecological footprint/biocapacity (%)
Arable products	126
Pasture grass	145
Wood	102
Fish	170
Total	**121**

other economic sectors; the country makes significant contributions to supplying the world with industrial products.

But if any country were to observe "Overshoot Day" with regard to its ecological footprint, it would be the "Middle Kingdom."

12

Species Extinction and Biodiversity

Introduction

"Mankind is causing the sixth mass extinction in Earth's history."

You have probably heard or read this or a similar headline at some point in recent years. Species extinction/biodiversity loss is one of the most prominent and present ecological issues of the present day, along with climate change, depletion of resources, loss of rainforests, and possibly plastic waste in the oceans.

But guess what percentage of vertebrates have become extinct in the last 500 years? Is this

- 3%
- 10%
- more than 10%?

All wrong – it is only about 1%. So, not for the first time in this book, the question arises: how do headlines/public perception and the facts fit together?

Let us embark, then, on a short but exciting journey into the complex realm of animals and see how they have evolved over the last few centuries.

© Springer-Verlag GmbH Germany, part of Springer Nature 2022

T. Unnerstall, *Factfulness Sustainability*,
https://doi.org/10.1007/978-3-662-65558-0_12

Within the living creatures on our planet, four major categories are distinguished:

- Vertebrates – in the five so-called "classes": mammals, birds, fish, reptiles and amphibians.
- Invertebrates – insects, worms, mussels, snails, etc.
- Plants
- Other organisms – fungi, viruses, etc.

Reliable scientific data are available for only one of these categories, namely for the approximately 75,000–80,000 vertebrate species. The other three categories are much less well researched so far. In particular, there are still very divergent estimates of the total number of species; however, there is some certainty that the number exceeds 10 million. The number of insect species alone is currently estimated to be at least 7 million. To make matters worse, there is intense scientific debate about how the term "species" should even be meaningfully defined in this category of animals.

It is obvious against this background that all statements about species diversity in these three categories, about species extinction, about the decline of populations etc. – at least on a global scale – are fraught with great uncertainties. For this reason, with regard to the topic of biodiversity and species extinction, I must confine myself to *vertebrates*.

The vertebrate kingdom as a whole is already fairly well researched, but human knowledge continuously advances here as well. In the last 20 years alone, more than 10,000 new vertebrate species have been discovered, and more are suspected (cf. Table 12.1).

Thus, about 10% of the currently existing vertebrate species are not yet discovered/described according to today's estimates. The data situation is worst for amphibians (frogs, newts, salamanders, etc.).

Table 12.1 Number of species in vertebrates; figures rounded. (Source: Wikipedia)

Vertebrate class	Known species	Further suspected species
Mammals	6000	0 to few
Birds	11,000	0 to few
Fish	34,000	1000–6000
Reptiles	11,000	0 to few
Amphibians	8000	2000–7000
Total	**70,000**	**5000–10,000**

Like all life on earth, vertebrates have undergone tremendous changes and upheavals in the course of earth's history after their emergence about 500 million years ago. On the one hand, there were five so-called "*mass extinctions.*"[1] Mass extinctions are defined as relatively short periods (probably a few 10,000 years) in which extreme external influences – volcanic eruptions, meteorite impacts, drastic climate fluctuations – caused the extinction of the majority of animal and plant species: at least 50%, in some cases even more than 70% of all species.

Besides these extreme periods, there is also the so-called "natural extinction rate." This means that species come and go in the course of evolution anyway, even without extraordinary external influences. They disappear, for example, by splitting into two new species, by mixing species or due to competition with other species. How long a vertebrate species exists on average in the course of evolution – its "life expectancy," so to speak – is a scientifically disputed question. However, most current estimates agree on 0.5–1 million years.

To put it another way: out of 1 million species, one to two species die out every year on average (and about the

[1] The five major mass extinctions are dated to 444 million years, 372 million years, 252 million years, 201 million years and 66 million years before our era. The exact causes of these events have not yet been fully clarified scientifically.

same number of new species appear). In concrete terms, this means two things:

- On average, about one vertebrate species dies out naturally every decade.
- Most of today's vertebrate species did not even exist 1 million years ago, and since the last mass extinction 66 million years ago (to which the dinosaurs fell victim), the spectrum of vertebrate species has already changed completely many times.

Against this background of the "normal speed" of evolution, the question now arises: How big is the influence of man on species extinction and biological diversity (= biodiversity)?.

One can distinguish here between two issues:

1. *Extinction of species* (biodiversity in the sense of the number of species)
2. *Decline of animal populations* (biodiversity in the sense of number of animals within each species).

Of course, both aspects are not completely independent of each other, since a continued decline in the sense of (2) will lead to the endangerment of a species in the sense of (1). But they are still two different issues, and will thus be treated separately.

Extinction of Species

The Facts – Global View

Massive human interventions in ecosystems and especially in vertebrate biodiversity are not new. When our ancestors reached Australia about 40,000 years ago, almost all of the 24 large animal species living there disappeared within a few millennia. The reasons are not completely undisputed, but most experts assume that humans hunted these animals to extinction as a source of food.

However, the focus here is on modern times. Since the year 1500, the extinction of species has been quite well researched (often retrospectively via fossil finds), and all relevant findings are compiled in the famous "Red List" of the IUCN.[2] All following data are taken from this IUCN database.

Of the known 70,000 (vertebrate) animal species, 0.5% have definitely become extinct in the last 500 years and another 0.5% are considered possibly extinct because no specimen of the species has been sighted for many years. Assuming a pessimistic point of view and summarizing both categories as "extinct," Table 12.2 results.

What species are we talking about here? There are a few prominent examples, such as the aurochs (extirpated in

Table 12.2 Number of vertebrate species extinct since 1500, world. (Source: IUCN)

Vertebrate-Class	Number of species	Definitely extinct	Possibly ext.	Total	% of all species
Mammals	6000	85	25	110	1.8
Birds	11,000	160	20	180	1.6
Fish	34,000	80	90	170[a]	0.5
Reptiles	11,000	30	40	70	0.6
Amphibians	8000	35	135	170	2.1
Total	**70,000**	**370**	**330**	**700**	**1**

[a]Ninety percent of the extinct fish species are freshwater fishes

[2]The IUCN (= International Union for Conservation of Nature) is an international non-governmental organization (and also the umbrella organization of numerous international governmental and non-governmental organizations) based in Switzerland.

1627), Steller's manatee (1786), or the giant bird Alk (1852), which were originally relatively widespread. However, these examples are atypical: in most cases, extinct species are animals whose habitat was limited from the start to small islands, individual lakes, or very specific areas of land, and thus could not escape when disturbed by human civilisation.[3]

Of great importance, of course, is the question of whether species extinction has accelerated in recent decades – that is, what the development over time of human-caused extinction looks like. This question is answered in Fig. 12.1.

For mammals, birds, fish, and reptiles, the rate of species extinction has slowed significantly over the past 30 years (even if all those on the IUCN list as possibly extinct are considered to be actually extinct), at least certainly not accelerated.[4]

The data situation is different for amphibians: here there seems to be no easing trend. This is not surprising: amphibians have their habitat at rivers and lakes, and exactly these habitats are increasingly claimed by humans in the course of the growing world population.

Even greater attention in science and also in the media is given to *species threatened with extinction*. In the IUCN system, there are three levels of threat to a species: vulnerable, endangered, and critically endangered. With respect to the

[3] A prominent example is Lake Victoria in Africa. Here, about 50 of the ca. 250 endemic fish species became extinct as a result of the introduction of the Nile perch by humans, whose food competition they were unable to cope with.

[4] Definitely extinct according to the IUCN list since 1990 are: four mammal species, eight bird species, four fish species, one reptile species, and one amphibian species. In the case of mammals, these are two rodent species, one bat on a small island in the Indian Ocean, and the Chinese river dolphin.

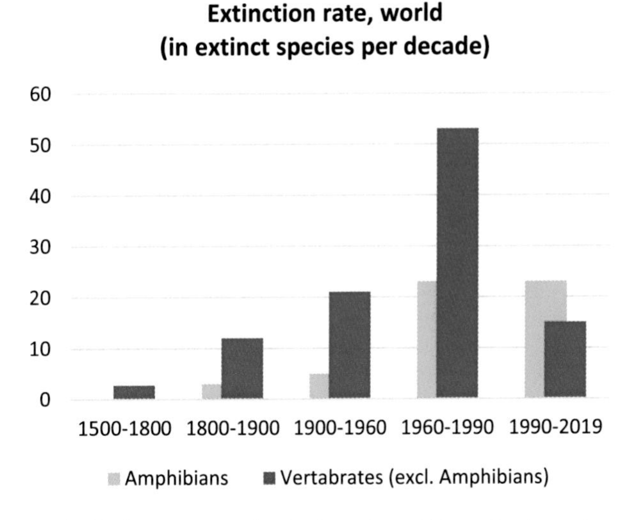

Fig. 12.1 Extinction rate (extinct and possibly extinct species per decade) in the world, 1500–2019. (Source: IUCN)

two upper levels, the current situation is as shown in Fig. 12.2.[5]

Except for the amphibians, about 1100 of the then 60,000 species are considered to be extremely endangered and about 2200 are considered to be endangered.[6] In the case of amphibians, the situation is again more critical.

An analysis of the temporal development of the endangerment situation over longer periods of time is hardly possible due to the lack of data. The only possible assertion is

[5] It should be mentioned in this context that of the 34,000 fish species described, only a good half have even been classified in terms of their endangerment; for the other half, there is insufficient data to assess the endangerment situation. Therefore, the number of endangered/extremely endangered species is often (and in Fig. 12.2) only related to the classified species; then a total of about 3% of all (vertebrate) species come out as extremely endangered and 6% as endangered. The often read statement "x percent of all animals are endangered" assumes that among the not classified, less known species just as many are endangered as among the classified species. This is hardly permissible, at least from a strictly scientific point of view.

[6] An additional 3000 species are classified as vulnerable.

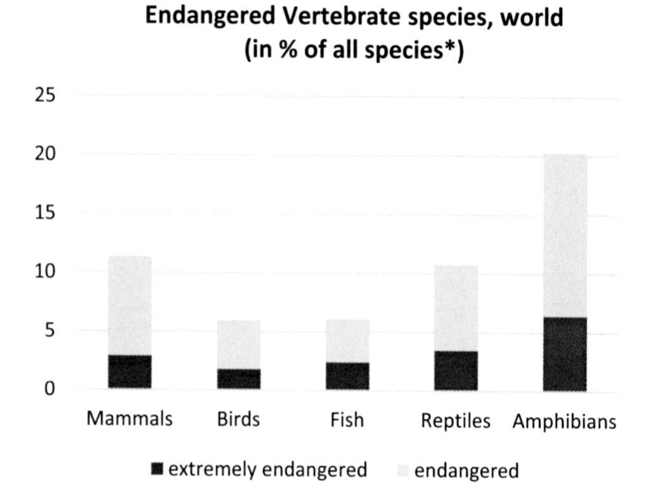

Fig. 12.2 Endangered vertebrate species of the world by class, in % of known and IUCN classified species, 2019 data. (Source: IUCN)

that in the last decades the list of endangered or extremely endangered species has grown by about 100 species.[7]

The Facts – The West

The West – here Europe and North America[8] – is home to about 10,000 species of vertebrates (i.e. about 15% of all vertebrate species on earth). According to the IUCN, 38 (= 0.4%) of these species have become extinct in the last 500 years, mostly (freshwater) fish species (see Table 12.3).

[7] This is a net consideration: de facto, 130 species have been added to these categories, but 30 species have also been removed because their situation has improved.

[8] In the data we present here for Europe and North America, Hawaii and the Canary Islands are not included, because – as isolated islands far from the continents – the situation there is very different from the one on the continents. For example, of the original endemic 70 bird species on Hawaii, 23 are now extinct and most of the other species are considered endangered.

Table 12.3 Number of vertebrate species[a] extinct in the West since 1500. (Source: IUCN)

Vertebrate class	Number of species	Extinct since 1500	In % of all species
Mammals	900	3	0.3
Birds	1900	4	0.2
Fish	6900	31	0.5
Reptiles	500	0	0
Amphibians	400	0	0
Total	**10,000**	**38**	**0.4%**

[a]The seven extinct mammal and bird species in Europe and North America are: Aurochs (1627), lake mink (1894), Sardinian hare (c. 1800), giant eel (1852), Labrador duck (1878), passenger pigeon (1914), Carolina minnow (1918)

No mammals and no birds have become extinct in the last 100 years, but about 25 fish species have; since 1990, all species have been conserved.[9]

Looking at the endangered species, ca. 5% of vertebrate species in the West are considered extremely endangered or threatened[10], see Fig. 12.3.

These relatively low numbers are probably due to the fact that in Europe and North America in particular, considerable efforts have been made for decades to protect threatened species. There are, often accompanied scientifically, a multitude of breeding programs, reintroduction attempts, new/enlarged nature reserves/national parks, etc.

Certainly, the above figures are no reason to give the all-clear, but biodiversity in the West is relatively stable at the moment in terms of the number of species. You will see below that the same is true for the second relevant issue, the question of animal populations.

[9] A few fish species whose extinction dates cannot be accurately determined may also have become extinct after 1990.

[10] Another 6% are considered vulnerable compared to 7.5% at the world scale.

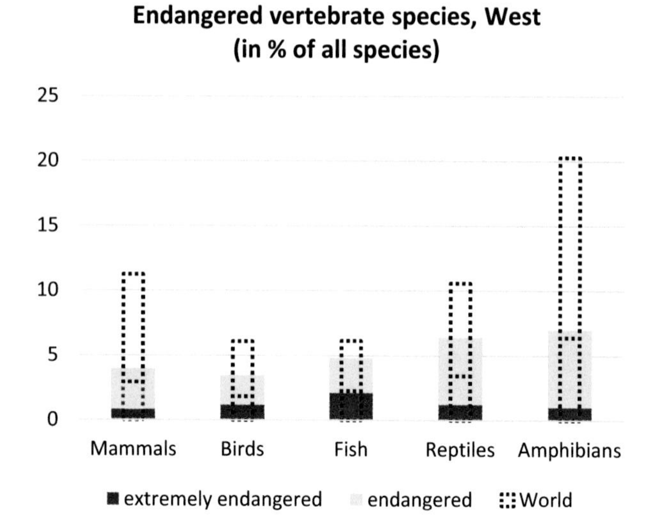

Fig. 12.3 Endangered vertebrate species in Europe (excl. Canary Islands) and North America (excl. Hawaii) by class, in % of known species classified by IUCN (Cf. footnote 5), 2019 data. (Source: IUCN)

Evaluation/Projection

Again and again, man has intervened deeply in nature from the beginning: He has deforested individual stretches of land (such as the Mediterranean region), he has exterminated large animals such as the Australian diprotodon, the mammoth, or the aurochs, and he has created large fields with only one arable plant where previously there was a colorful diversity of plants. By and large, however, until modern times there were sufficiently large and undisturbed habitats for the vast majority of animal and plant species to survive; in particular, for most mammals, birds, fish, reptiles, and amphibians.

In recent centuries, as modern civilization has spread to even the most remote areas and islands, and as the world's population has grown, this picture has changed significantly. An increasing number of vertebrate species are threatened in their existence. This also has to do with a phenomenon of evolution that most of us are probably not aware of. The typical vertebrate species is not, for example, the lion, the golden eagle or the tiger shark – both distributed over thousands of kilometers and at home in many regions. Rather, the typical species is the Christmas Island pygmy bat, which precisely only occurred on the tiny Christmas Island and is the latest entry in the list of extinct mammal species (2017). Indeed, most vertebrate species have small habitats,[11] are exclusively adapted to the environmental conditions there, and are therefore very sensitive to even minor disturbances of these conditions by human civilization or climate change.

Fortunately, the threat to vertebrate species posed by human civilization has, for the most part, not yet had irreversible consequences; 99% of the more than 70,000 species that existed 500 years ago still exist today. Nevertheless, during these last 500 years, an average of about 10 vertebrate species (excluding amphibians) have become extinct per decade. This rate has increased to over 30 in the twentieth century, and currently it is at about 15. This exceeds the evolutionary natural extinction rate (of about 1 species per decade) by a factor of 15. In the case of amphibians, this "acceleration factor" has even exceeded 200 in recent decades. This is the reason why many scientists speak of a "6th mass extinction" that humans are about to cause; and the media have readily adopted this verdict in their headlines.

However, as plausible as this may seem at first glance, it appears dubious upon second thought. If the above figures are taken as a basis, the human-caused extinction of

[11] The same is true to an even more extreme degree for insects. Researchers believe that there are insect species found on only one tree(!) in the rainforest.

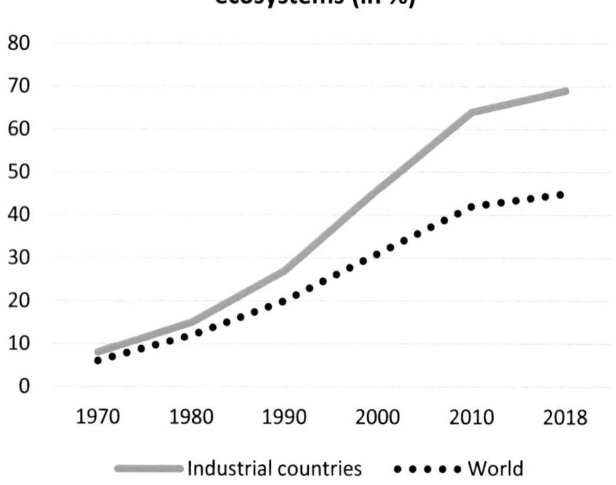

Fig. 12.4 Quota of ecosystems of special value under protection according to IPBES, 1970–2018. (Source: IPBES 2019)

mammals, birds, fish and reptiles would have to continue at exactly the same rate for another 20,000 years to reach the threshold of 50%[12] of disappeared species, so as to speak of a "mass extinction".

Given the utter impossibility of predicting the development of human civilization even in broad terms for the next 500 years, such an extrapolation seems to me highly questionable. All the more since mankind does not passively watch the threat to many animal species – on the contrary. It is not only the economy and consumption that have grown in many regions of the world over the past decades, but also ecological awareness and efforts to preserve endangered animal species. An impressive illustration of this is Fig. 12.4.

[12] Even in the case of amphibians, it is more than 5000 years.

In the most developed countries, 70% of the ecologically particularly sensitive/valuable areas have already been placed under protection, and the trend is intact. The success is tangible: species extinction has been largely halted in the West in recent decades.

Of course, no one can rule out the possibility that further societal development in Europe and North America – and in the world anyway – will again lead to a significant deterioration in conditions for wildlife and for endangered species in particular. But either way, extrapolating from today's state of affairs to many millennia of future human history is simply not reasonable. The headline "Mankind is causing the 6th mass species extinction in Earth's history" therefore evokes, in my opinion, a rather distorted view of the real ecological situation.

Another very relevant issue in this context is the question of the future impact of **climate change** on species extinction. To what extent can animal and plant species migrate with the expected shift in climate zones, to what extent can they adapt evolutionarily? This question is the subject of intensive biological research and discussion and, of course, depends crucially on the scenario taken as a basis, i.e. the extent of climate change. A basic study from 2015[13] predicts that in a medium scenario (temperature increase 2.5–3 degrees by 2100), about 6–8% of species could become extinct by 2100. Here, the risks are particularly high for South America (many species with small habitats), Australia and New Zealand (hardly any migratory movements possible), and amphibians are again the most affected vertebrate class.

According to this forecast, then, climate change almost certainly represents the most serious threat to living conditions for animal and plant life caused by mankind.

[13] Urban (2015).

Decline of Animal Populations

The Facts

The IUCN Red List is the standard source for all data on species diversity. The World Wildlife Fund's Living Planet Index (LPI) has a similar goal for tracking animal populations on Earth. The LPI is limited to vertebrates by design, i.e., it does not cover insects and others. Since 1970, all available observation and measurement results concerning the number of animals in certain vertebrate species have been recorded and evaluated in this database on a scientific basis.

Compared to the "Red List," however, the data must be characterized as very thin: The LPI monitors only about 4500 of the currently known 70,000 vertebrate species (i.e. less than 7%), and for each of these species only a part of the existing populations.

On this narrow basis, a very complex methodology is then applied to extrapolate the observations to all species and thus derive statements on the overall development of animal populations in mammals, birds, fish, reptiles and amphibians – both in individual world regions and as a global average.

It is obvious that the quantitative results can only claim a rather limited reliability – but some qualitative trends can be established with a resonable degree of certainty.

Trend 1

Animal populations have declined significantly on a global scale since 1970.[14] Most of these declines occurred in the

[14] The actually observed decline is only on the order of 20%; it is only by the LPI-methodology that the numbers of Fig.12.5 are derived.

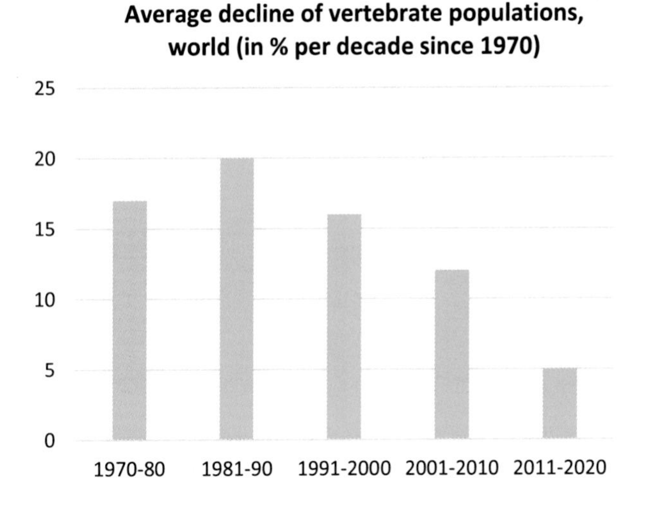

Fig. 12.5 Average decline of vertebrate populations in the world, 1970–2020 (2011–2020 extrapolated from 2011–2016 base), in % per decade. (Source: LPI)

1970s and 1980s; ever since 2000, the rate of loss has been greatly reduced (see Fig. 12.5).

Trend 2

Mammals, birds, reptiles, and marine fishes have been much less affected overall than freshwater species, i.e., freshwater fishes and amphibians (Fig. 12.6).

Trend 3

This is especially true for animal populations in protected areas. Here, mammals, birds and marine fish have actually rebounded, while amphibian populations continue to decline.

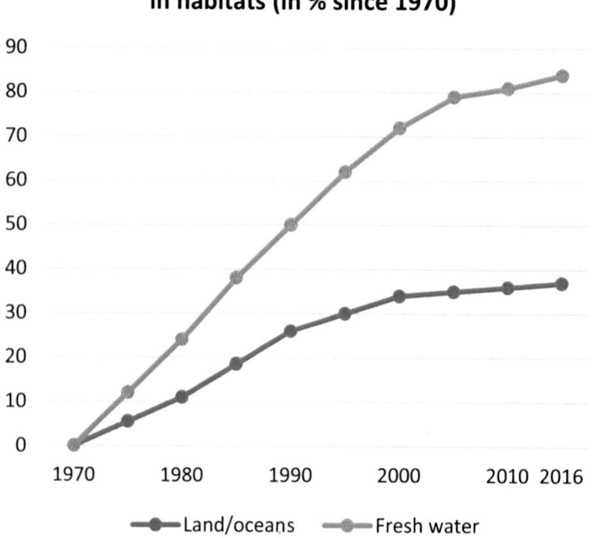

Fig. 12.6 Average decline in vertebrate populations in the world by living range, % since 1970. (Source: LPI)

Trend 4

The situation is clearly better in the northern hemisphere of the world than in the southern hemisphere. In North America, Europe, Russia, large parts of Asia and also the Arctic, animal populations have been declining moderately now for 10–20 years and are on average at a level of 70% of the 1970 numbers. In the southern hemisphere, they have declined more. In South America and the Indo-Pacific region in particular, populations seem to have shrunk by more than half on average (Fig. 12.7).

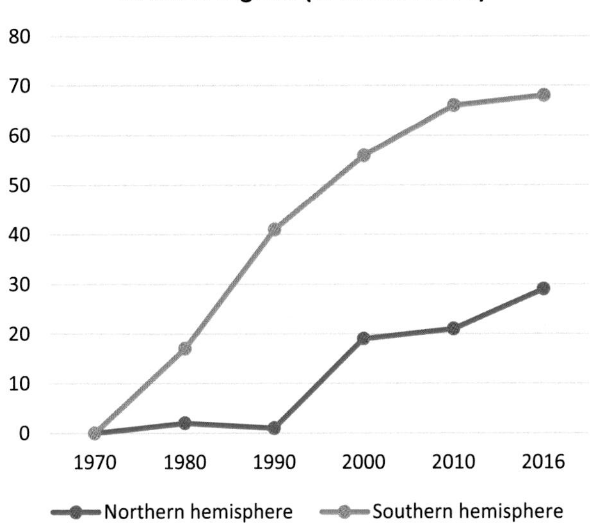

Fig. 12.7 Average decline of vertebrate populations in the world, by world region, in % since 1970. (Source: LPI)

Evaluation/Projection

The Living Planet Index reinforces the picture we have already gained on the subject of species extinction. After a phase of fairly reckless human expansion in the twentieth century, there has been a marked stabilization trend in recent decades – not only in terms of species diversity, but also in the size of animal populations. An exception to this trend is amphibians (and to some extent freshwater fish), whose living conditions continue to deteriorate in many regions of the world.

In the economically more developed countries, and especially in the West, the trend of declining animal numbers has been halted, in some cases even reversed, and species extinction has thus also been largely halted.

These positive developments – at least among vertebrates, i.e. the higher evolved animals – are by no means a guarantee for the future. But they do show impressively that growing environmental awareness, the worldwide efforts to protect biologically particularly sensitive ecosystems, and many individual species protection measures around the globe are bearing fruit. And they show that economic progress and the peaceful coexistence of humans and animals are not only not contradictory, but can even go hand in hand.

Excursion: Overfishing the Oceans?

"More than 90% of all global fish stocks are fully or overexploited". This statement can be found in countless headlines, on the websites of environmental organizations, the World Economic Forum and many others. To substantiate this claim, the same data from the FAO are quoted again and again[15] – without, however, making the effort to understand how these figures are collected and in which context they are to be interpreted. The topic of "overfishing the oceans" is therefore a good example of how misleading it can be to interpret figures detached from the overall context.

In detail:

- There are 35,000–40,000 fish species worldwide. Only about 1600 species (4–5%) of these are commercially fished, and about 200 species account for two-thirds of

[15] FAO (2018).

the total fish catch. All published figures concerning fishing are only about these relatively few fish species, not about "all fish stocks in the oceans."

- The cited FAO data are not about fish species, but about so-called "fish stocks." A fish stock is a larger group of fish in a marine area that is largely isolated from other groups of fish and therefore biologically self-sufficient, i.e., it lives and reproduces only within that area. It may include one or more species of fish, and one species may be present in a whole range of fish stocks.
- How many such "fish stocks" there are in the world is unknown; but about 1500 of these areas are regularly fished, so they form the basis of commercial global fisheries. Again, a portion of these (about 500 fish stocks) are regularly monitored scientifically, and only these results are included in FAO statistics.
- In its latest report of 2018, FAO concludes on this basis that two-thirds of the monitored fish stocks are sustainably fished: Only the amount of fish is caught that corresponds to the natural reproduction rate, i.e., the number of fish remains constant. This situation can certainly also be described with the vocabulary "fully exploited" or "fished to the biological limit" – but this does not change the fact that this is a *sustainable* form of exploitation.
- One-third of the fish stocks surveyed (about 170) have been or are currently being overfished, meaning that fish numbers are declining and the ecosystem is being damaged. The longer-term consequences are primarily commercial: as fishing in the area becomes increasingly unprofitable, fishermen often have to move to other areas. This gives the ecosystem a chance to regenerate, i.e. fish numbers grow again and approach the original stock in

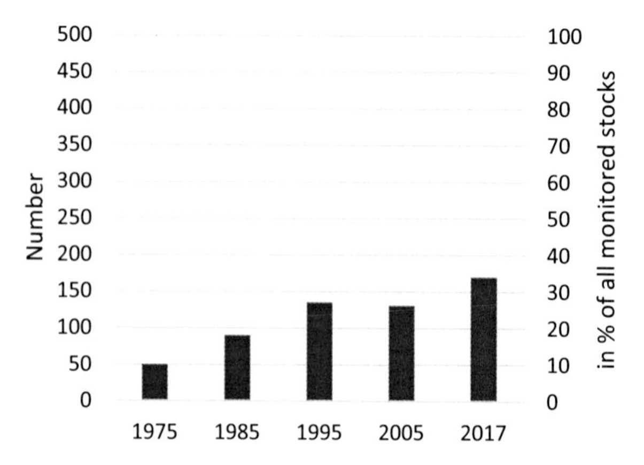

Fig. 12.8 Number of unsustainably exploited fish stocks in the world, 1975–2017. (Source: FAO 2018)

the longer term. Therefore, unsustainable exploitation of a fish stock does not usually pose a threat to the survival of a fish species – especially since most species are found in multiple fish stocks. The number of these overexploited areas is increasing, but moderately so: between 1995 and 2017 (the last published year), about 35 more areas were added, see Fig. 12.8.

- Global fish catch has stagnated at about 90–100 million tons of fish per year for about 20 years. The global increase in total fish production is therefore exclusively brought about by aquaculture.[16] cf. Fig. 12.9.

[16] This trend is very likely to intensify: just as meat production is now based on livestock rather than hunting in the wild, the future of fish supply lies in aquaculture rather than fishing in the oceans.

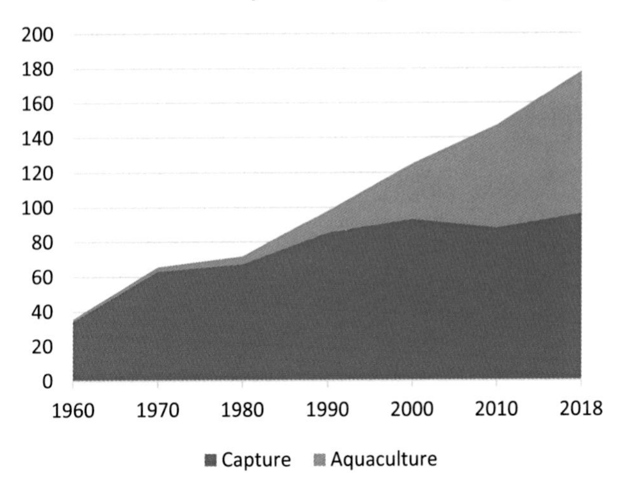

Fig. 12.9 Global fish production (including other marine species), 1960–2018, in million tons. (Source: FAO 2020)

Conclusion

There can be no question at all of "the oceans being fished empty": first, most fish species are not affected by human fishery; second, for many of the fish species affected, only a smaller proportion of the populations is even monitored; and third, the vast majority of the populations/fish stocks monitored are sustainably fished, i.e. not depleted.

This does not mean that the overexploitation of fishing grounds is not a problem and should not be stopped – also in the interest of the respective local fishing companies, which otherwise deprive themselves of their own income basis.

13

Forest Loss – Deforestation of Rainforests

Introduction

"Forest" – this word evokes positive associations in most of us. We associate it with tranquility, recreation, nature, or even – especially in the specification "tropical rainforest" – with exoticism, fascinating flora and fauna, adventure. This emotional connection to the forest is no wonder. For millennia, forests have been crucial to man's civilizational development: not only as a supplier of firewood for cooking and heating, but also as a source of building materials for houses, transportation, irrigation systems, fences for livestock, and much more. Without tree trunks, man would not have been able to reach Australia 40,000 years ago, nor would the Romans have been able to build their empire. Columbus' "Santa Maria" 500 years ago was a pure wooden ship, and even today almost everyone on earth is surrounded in everyday life by furniture whose material grew in a forest.[1]

---

[1] In addition to the many traditional functions that forests have for humans, there is another one that has only come into greater focus in recent decades, primarily from a scientific perspective. It is estimated that tropical rainforests harbor 70–80% of all animal and plant species on the planet, and thus they contain an immeasurable gene pool that could be important for future medical developments.

© Springer-Verlag GmbH Germany, part of Springer Nature 2022

T. Unnerstall, *Factfulness Sustainability*,
https://doi.org/10.1007/978-3-662-65558-0_13

Against this background, it is understandable that 40 years ago the term "Waldsterben" hit Germany like a bomb; and that we are particularly moved by images of centuries-old trees falling victim to large saws. For some years now the forest has once again come upfront in politics and media, but this time for a different reason. In the increasing worldwide discussion about climate change and its causes, the global loss of forests and especially the deforestation of tropical rainforests is one of the central topics.

Indeed, an estimated 30% of human CO_2 emissions are absorbed by the world's forests, and conversely, when forests are cleared, significant amounts of the carbon stored in the trees are often released into the atmosphere, contributing to climate change.

So what is the state of the Earth's forests, what is the truth in headlines such as "The Amazon rainforest is disappearing," and what developments can we expect in the coming decades?

The Facts – Global View

The forest is – since its "invention" by evolution about 300 million years ago – the dominant ecosystem on the continents of our planet. If we exclude the deserts, icy areas and high mountains, more than half of the land area was originally covered with forest[2]: about 60 million km^2. In Central Europe, it was even up to 90%.[3] In the millenia until 100 years ago, man cleared about 10 million km^2 – to gain land for agriculture, and to get a hold on building material.

[2] By "original" we mean here: about 10,000 years ago, after the forest in the northern hemisphere had "reclaimed" the areas covered with ice during the last ice age.

[3] In one of the first reliable historical sources about Germany, the Roman historian Tacitus (about 100 AD) refers to Germania as the "land of terrible forests."

Table 13.1 Forest area of the world in million km²

	Temperate forest	Tropical forest	Total
8000 BC	≈30.0	≈30.0	≈60
1920	≈22.5	≈27.5	≈50
2020	22,2	18,4	40,6

Source: FAO 2012, 2015, 2020

The main areas of deforestation were in the more densely populated temperate latitudes – i.e. in Europe, North America and East Asia (Table 13.1).

In the last 100 years, another 10 million km² (= 1 billion ha) of forest were cleared, but this time almost exclusively in the tropics.[4] The temporal development over these 100 years is shown in Fig. 13.1.

It shows that the global forest loss outside the tropics has stopped: The forest area in the temperate zones of the earth has actually been increasing again since 1990. The most important reason for this is the massive afforestation of forests in China. In the tropics, on the other hand, forest loss is continuing, but here, too, a slowdown in the rate of deforestation [5] can be observed (cf. Fig. 13.2).

As can be seen, 2–2,5% of tropical forests are currently being lost per decade, measured against the original stock.

Particularly in the focus of ecological concerns is the part of the tropical forests commonly referred to as **"tropical rainforest."** This particular form of vegetation is characterized by

[4] "Tropics" refers to the land areas within a ± 2500 km strip around the equator – about 50% (50 million km²) of the land areas excluding ice areas, deserts, and high mountains.

[5] A distinction must be made between the *rate of forest loss* and the *rate of tree cover loss*, which is recorded, for example, in the Global Forest Watch. The latter includes, in particular, fires that do not completely destroy the forest (it recovers over the years), but lead to degradation and reduced canopy cover. This rate has been unusually high in recent years (due to increased fire events).

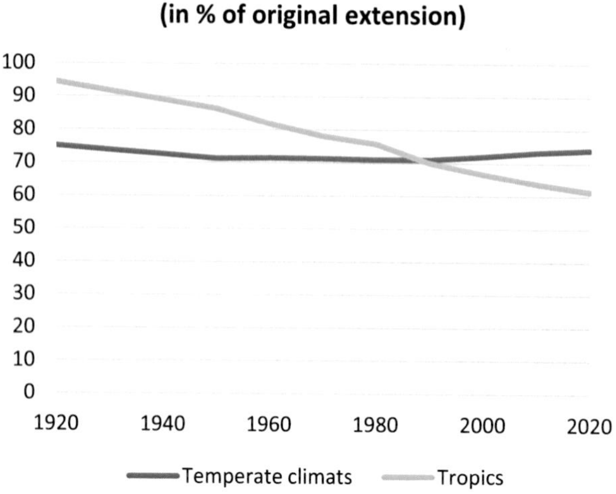

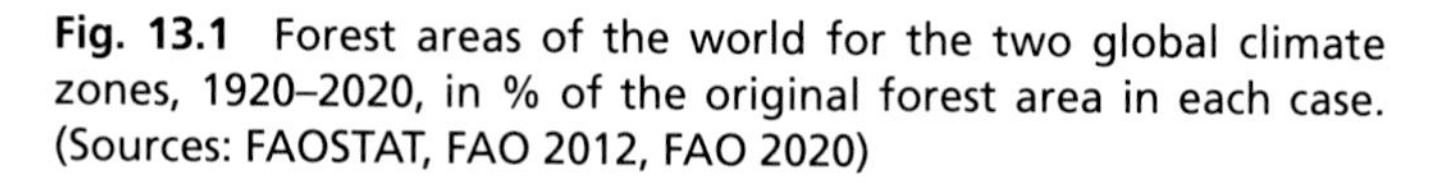
Fig. 13.1 Forest areas of the world for the two global climate zones, 1920–2020, in % of the original forest area in each case. (Sources: FAOSTAT, FAO 2012, FAO 2020)

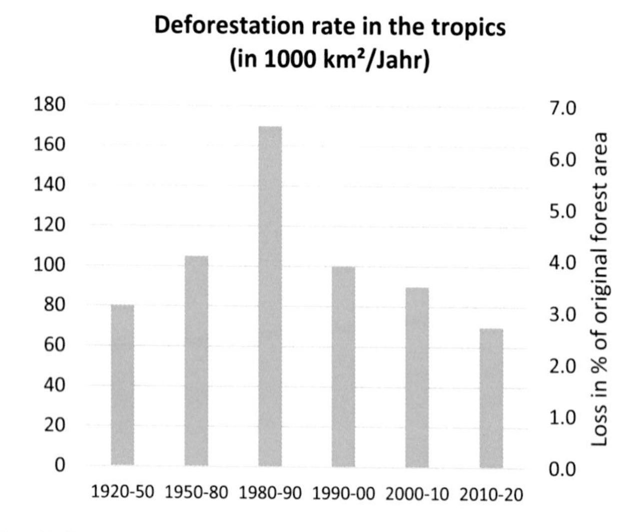

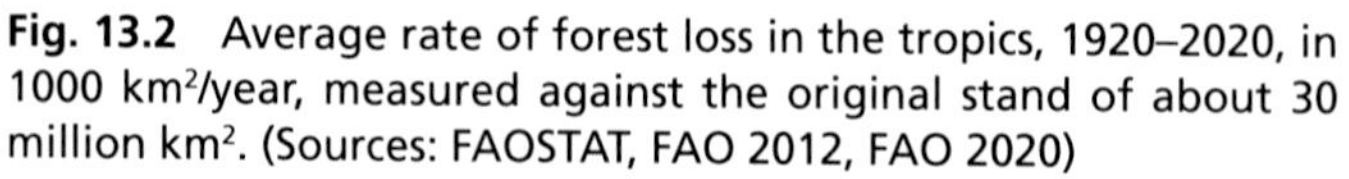
Fig. 13.2 Average rate of forest loss in the tropics, 1920–2020, in 1000 km^2/year, measured against the original stand of about 30 million km^2. (Sources: FAOSTAT, FAO 2012, FAO 2020)

- very high precipitation (about five times as much as in Central Europe),
- consistently high temperatures of 25–30 degrees, thus no seasons,
- extremely numerous, but also extremely specialized, animal and plant species that have very small habitats (see footnote 11, Chap. 12) and very small population sizes,
- intensive material cycles on several levels, but mostly rather infertile soils. Therefore, unlike other forms of forest, rainforest cannot be replanted, and its natural formation takes 50–100 years.

Tropical rainforests are also often referred to as the Earth's "green lungs" – they would produce 20% of atmospheric oxygen. This is a false idea, but one that persists in many people's mind.[6] The rainforests have a largely even oxygen balance.[7]

What is the state of tropical rainforests?

An important difficulty in answering this question is the fact that the term "tropical rainforest" is not precisely defined, and therefore a whole range of different figures circulate. Depending on which forests are included, there are 6, 9 or 11 million km^2 of tropical rainforest left today (out of a total of about 18 million km^2 of tropical forests) and correspondingly different rates of loss are reported. In the following, I use the FAO's term "primary tropical forest," for which there are longer, consistent data series.

The development of forest loss for "primary tropical forest" shows a similar pattern as for tropical forest as a whole

[6] "The Amazon rain forest – the lungs which produce 20% of our planet's oxygen – is on fire," said Emmanuel Macron on Twitter on August 22, 2019 (@emmanuelMacron).

[7] For a good explanation, see e.g. Zimmer (2019).

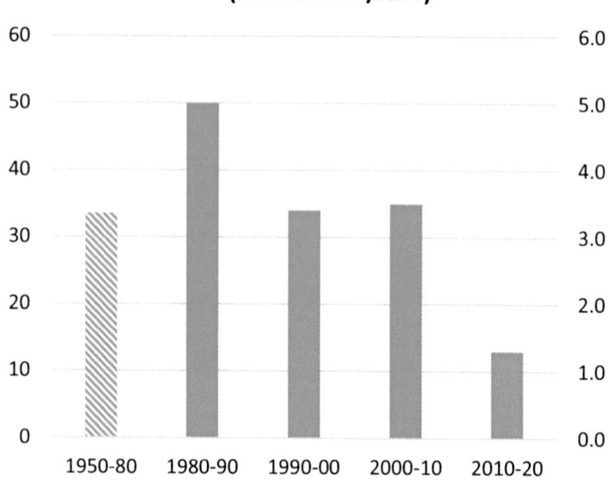

Fig. 13.3 Average rate of forest loss in "tropical primary forest," 1950–2020, in 1000 km²/year (pre-1980 data is uncertain). (Sources: FAOSTAT, FAO 2012, FAO 2020)

(Fig. 13.3): after a peak in the 1980s, the destruction of these original rainforests has declined significantly.

Currently, about 5 Mio. km² are classified as primary forest in the tropics, and 10,000–15,000 km² per year of this tropical jungle are victim of human activities.

The Facts – The West

The history of forests in the West – in this case, Europe (excluding Russia) and the United States – is quickly told. Originally, about 70% of Europe was covered with forest, but significant portions of these woods disappeared well before modern times with the spread of civilisation, when they fell victim to either settlements, cropland, pasture, or

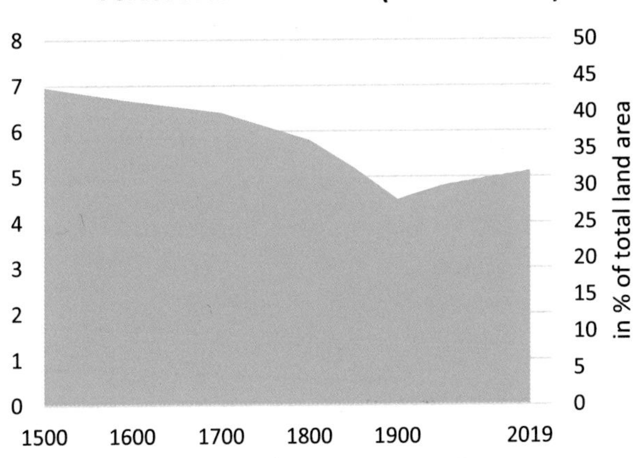

Fig. 13.4 Forest area in Europe (excluding Russia) and the United States, 1500–2019. (Sources: OurWorldinData (Forests), FAOSTAT)

the need for building materials. There was a further surge in deforestation in the nineteenth century in the wake of the Industrial Revolution, so that by 1900 forest areas had shrunk to less than 30% of the land area. Since then, they have been slowly increasing again.

In the USA, the forest was largely untouched before the arrival of the Europeans, but then experienced a very similar development as on this side of the Atlantic: there, too, the low point was reached at the beginning of the twentieth century, since when there has been a positive trend, see Fig. 13.4.

Projection Until 2050

The facts speak for themselves: In the planet's temperate climate zones, forests are slowly growing again, primarily through targeted reforestation. In the tropics, deforestation continues, but rates have been declining for about 30 years and have now reached a level of about 2–3% per decade. The same is true for the particularly valuable and ecologically sensitive tropical primary forests, which include in particular the untouched rainforests that are often the focus of attention.

Looking ahead to the next few decades, there is little to suggest that these trends will change fundamentally. *Outside the tropics*, afforestation is likely to increase in many regions (also to counter climate change)[8]; in particular since there is no need for additional cropland or pasture. Forest area here can therefore be expected to continue to grow (slowly).

In the tropics, there may well be temporary fluctuations in the efforts to conserve forests and, in particular, to further curb rainforest deforestation. But global public pressure on the responsible governments – primarily Brazil, Indonesia, and the Democratic Republic of Congo – in this direction is increasing, and therefore it is unlikely that the now long-standing positive trend will be completely reversed.[9] Consequently, it is a rather conservative estimate to extrapolate the figures of the last 10 years to 2050 (i.e., to assume no further reduction in deforestation rates).

On this basis, there should still be about 16 million km^2 of forest in the tropics in 2050 – compared to 18 million km^2 today and 30 million km^2 originally. Of today's tropical primary forests (about 5 million km^2), no more than 10% are expected to have disappeared by mid-century, leaving an area still standing larger than the EU.

[8] Global reforestation is seen as probably one of the most effective measures to produce "negative emissions" (i.e., actively remove CO_2 from the atmosphere) and thus help climate protection.

[9] Indeed, in Nov. 2021 at the climate summit in Glasgow, the main countries involved committed themselves to end rainforest deforestation as early as 2030 – an encouraging signal.

Evaluation/Summary

Deforestation of tropical rainforests is an issue that has been increasingly present in the media in many countries for decades. For a long time, the main concern was biodiversity, i.e. the disappearance of insect and plant species due to the destruction. For some years now, the fact has also come to the fore that this deforestation, primarily through the slash-and-burn clearing usually associated with it, causes a noticeable proportion of global CO_2-emissions (in the order of 2%).

Both concerns, the associated political discussions as well as civil protest actions have contributed to the fact that the overall situation today is better than usually conveyed by the media. To be sure, the current rate of deforestation means large areas in absolute terms (thousands of hectares of tropical primary forest are lost every day), and every square kilometer of rainforest destroyed is an irretrievable loss. On the other hand, the current rate of deforestation means that by 2050, more than 90% of the tropical rainforests that exist today will probably still be there.

Climate change could prove to be a greater threat to the rainforest within the next few decades. For example, there is increasing scientific evidence that even the current level of warming could have a lasting effect on natural processes in the Amazon region.[10]

Predicting the long-term future of rainforest deforestation, or tropical forests in general, is almost impossible – but at least it is not unlikely that the West will prove to be a role model in this regard. In Europe and the USA, the loss

[10] However, the exact extent of this is still a matter of debate, as is the thesis put forward by some climate scientists that there could be a "tipping point" in the area of rainforest in the near future – an area below which the natural cycle of rainfall-evaporation-cloud formation-rainfall over the Amazon basin could be interrupted, threatening the continued existence of the rainforest as a whole.

of forests over many centuries had the same reasons for which deforestation is taking place in the tropics today: Expansion of arable and pasture land, extraction of the coveted building material wood. But for more than 100 years now, this development has come to an end and the forest areas of the West are slowly increasing again.

In any case, it is clear that economic and civilizational development and the preservation of the unique ecosystem "tropical rainforest" are not mutually exclusive. Hope is justified – if the public, NGOs, politicians and many people on the ground continue to work on it – that this insight will translate into reality in the coming decades.

Excursion: "The Amazon Rainforest Is on Fire"

In the summer of 2019, an unusually high number of fires in the Amazon region is throwing the international community into a tizzy. The media are full of headlines that "the Amazon rainforest is burning"; French President Emmanuel Macron brings the issue to the agenda at the G7 summit; reproaches and protests are hailed against the Brazilian government.

On the one hand, this reaction can be considered positive: In an age of permanent satellite monitoring, of split-second information dissemination to all news agencies in the world, and of a public highly sensitized to environmental issues, no ecologically relevant event goes unnoticed. And every government must answer to the international community if it does not convincingly pursue (or if it even sponsors) alleged or actual environmental sins.

On the other hand, this is a good example of how such reports often lose sight of elementary aspects:

- The fires did not primarily affect intact rainforest, but rather areas that had already been cleared and dried out; thus, for the most part, they did not represent a loss of forest.
- Fires were more numerous than in previous years but were not of unusual proportions in view of recent decades. Overall, areas affected by fire worldwide have decreased in the last decades.[11]
- The affected area (around 1 million ha) represents only a fraction of 0.2% of the total rainforest area in the Amazon basin.

In short, the headline "The Amazon Rainforest is Burning" painted a false, or at least misleading, picture of reality. Such exaggerations certainly help in some cases to create pressure for action. In the long run, however, I think this kind of journalism / political discussion poses considerable risks to the credibility of media / politics and could lead to "fatigue" in public attention.

[11] See database globalfiredata.org

14

Plastic Waste in the Oceans

Introduction

If you had gone to a German bookstore in the summer of 2019 and looked for books on ecological issues, you would have found primarily – not books about energy transition and climate protection, but – books about plastic. "Plastic-free for beginners," "Save plastic," "Less plastic in the sea", "For an environment without plastic," these were the titles. At the same time, there were numerous discussions and reports on the pros and cons of plastic bags, on the EU ban on plastic straws, or even on Germany's exports of plastic waste to faraway countries.

The reasons for this wave of public attention that has been rolling around plastic for years boil down to one central point: mountains of plastic waste on many coasts, especially in Asia; gigantic floating islands of plastic parts in all oceans, the largest of which is more than four times the size of Germany. Plastic waste in the oceans: This is probably

© Springer-Verlag GmbH Germany, part of Springer Nature 2022

T. Unnerstall, *Factfulness Sustainability*,
https://doi.org/10.1007/978-3-662-65558-0_14

the main motive behind the numerous societal activities concerning plastic products in many countries.[1]

Indeed: everyone knows the pictures of dead seabirds, their stomachs full of plastic waste; the pictures of seals and sea turtles getting entangled in old fishing nets floating in the sea. And you may have read about the potential accumulation of small plastic particles (microplastics) in the food chain – from fish to humans. A few years ago, the grim reporting culminated in the headline[2]: "More plastic than fish in the ocean in 2050?"

There is no doubt about it: besides resource consumption, species extinction, rainforest deforestation and, of course, climate change, plastic waste in the oceans is one of the most prominent ecological issues. The widespread concern about the further development of this problem has moved many households to take a closer look at plastic in everyday life and their own options for action. This societal discussion and the commitment of so many people are undoubtedly positive. But it is also worth getting down to the objective facts and realistic forecasts for the coming decades. We will see that the picture that emerges holds some surprises.

[1] In this book, I focus on the classic visible plastic products and the resulting plastic waste. This is not to be confused with a completely different, but also increasingly discussed topic: the widespread distribution of so-called microplastics – very small plastic particles – in the environment and now also in the human body. They originate not only from decaying plastic waste, but also from tire wear, artificial turf, washing processes of synthetic textiles and many other sources. Since there are other causes and distribution mechanisms here, it is a separate problem that I – for lack of systematic global data – unfortunately cannot deal with in this book (but cf. the comments in the section "Evaluation/Summary").

[2] See, e.g., Süddeutsche Zeitung, Jan. 21, 2016.

The Facts – Global View

Plastics have had a phenomenal career; actually, they are still in the midst of it. Virtually unknown 70 years ago, they are now omnipresent in our daily lives, and the evolution of their global consumption far outstrips that of all other materials (see Fig. 14.1).

Why is that?

Plastics have many advantages: they are light, can be shaped as desired, are very durable, impermeable to air, waterproof, and inexpensive. They are therefore excellently suited for packaging, as materials in building construction, in medicine, for hygiene products, clothing, etc. But (like any material) they also have a disadvantage: unlike metals,

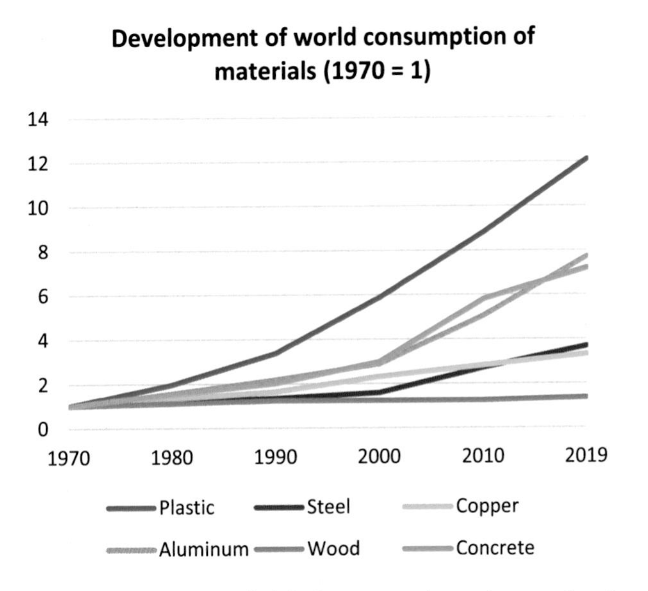

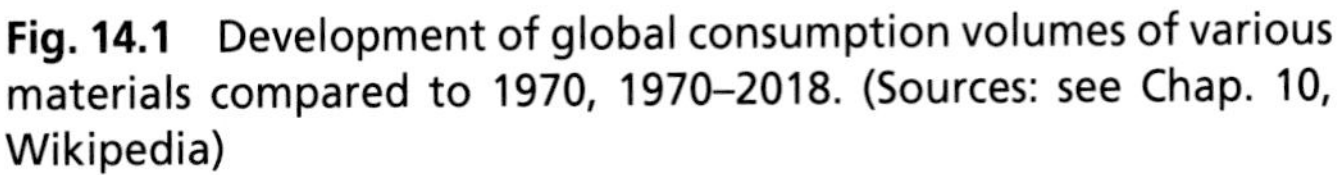

Fig. 14.1 Development of global consumption volumes of various materials compared to 1970, 1970–2018. (Sources: see Chap. 10, Wikipedia)

for example, they are hardly recyclable, at least with today's technologies. After one or two recycling cycles, they can only be incinerated or landfilled.

Therefore, the global volume balance of plastics produced to date looks quite different from that of metals (Chap. 10). From 1950 to 2020, about 10 billion tons of plastics were produced. Of this

- 3 billion tons are currently still in use (about 5% of them as recycled material);
- 1.2 billion tons have been incinerated;
- 4.3 billion tons have been properly disposed of in protected landfills;
- 1.5 billion tons were either stored in unprotected landfills or simply discarded/left somewhere.

The development of plastic waste over time is depicted in Fig. 14.2; it shows that currently about 30% of plastic waste worldwide is recovered, i.e. recycled or incinerated.

It is important to understand that properly landfilled plastic waste is not a real problem: plastics hardly react with the environment; they do not affect the surrounding air, soil, or groundwater.

The only real problem, therefore, is plastic waste (and indeed garbage in general) that is neither recycled in any way nor properly landfilled, i.e., not subject to proper waste treatment at all. For this plastic waste, two of the advantages of plastics, if you will, become disadvantages: They are light, so they can be blown or washed off the land into rivers or the sea; and they are very persistent, so they decompose very slowly.

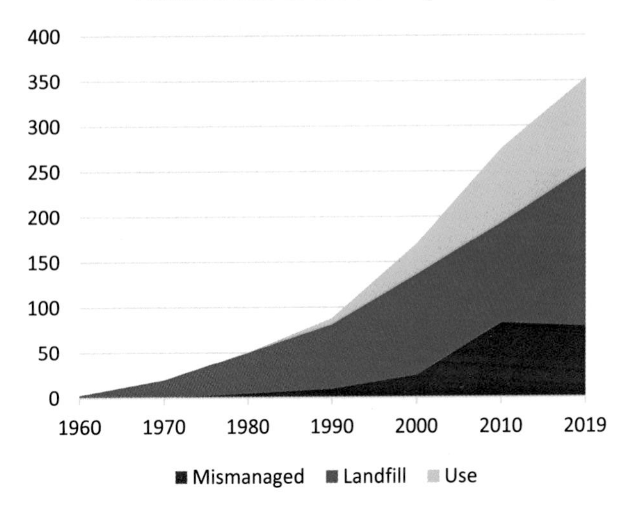

Fig. 14.2 World plastic waste by categories, 1960–2019, in million tons. (Sources: Geyer et al. 2017, OECD (2022), own calculations)

The **plastic waste in the oceans** actually originates precisely here: it stems for the most part[3] from plastic products discarded near the coast or stored in unprotected landfills, some of which are transported into the sea over time by wind and rain, either directly or by way of larger rivers.

How much plastic has accumulated in the oceans in this way over the last decades? This question was scientifically disputed for a long time, and until today one has to rely on rather rough estimates. A new, comprehensive study of the OECD concludes that of the ca. 1.5 billion tons of untreated plastic waste, 30 million tons have ended up in the sea (and a further 110 millions tons in rivers).

[3] A small portion (10–20%) also comes from ships that simply dump their trash, unused nets, etc. into the sea.

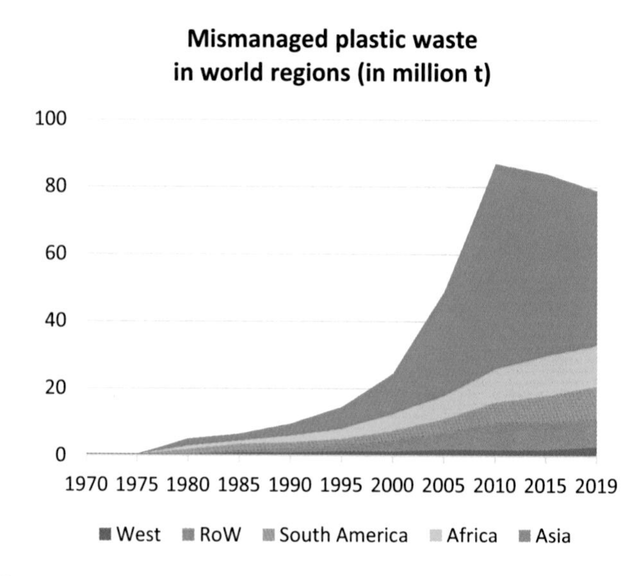

Fig. 14.3 Mismanaged plastic waste in four world regions, 1970–2019, in million tons. (Sources: Jambeck et al. 2015, World Bank 2019, OECD (2022), own calculations)

The core of the whole problem is therefore clearly a lack of proper waste treatment; and although plastic only accounts for 10% of total waste, in the case of plastic this shortcoming also shows up in the sea. Thus, the problem varies greatly from country to country, from region to region – depending on the level of development of waste management. Figure 14.3 shows the development over time of the mismanaged quantities of plastic waste, broken down by world region.

It is clear that over 70% of this waste was generated in Asia and Africa; correspondingly, over 70% of marine plastic debris originates from these regions.

The crucial question for assessing the ecological relevance of plastic waste in the oceans is now: what happens to this waste? Where have the many millions of tons of plastic gone? The plastic piles floating in the oceans are huge in terms of extent, but they have been accurately measured with the result that all together they account for only about 0.5 million tons, a small fraction of the estimated total.

The answer to this question is the subject of intense scientific research. The latest studies[4] come to the following conclusions:

- Over 70% of the seaborne plastic waste from the last decades is still located near the coast – i.e., on beaches, on the seabed a short distance from land, or floating in coastal waters.
- Only a small portion of the waste has drifted out into the open sea, where it forms, among other things, the visible plastic patches.
- About one third of the original quantities have now decomposed into so-called microplastics (plastic pieces <5 mm), which have either been ingested by marine organisms, are still floating in deeper water layers, or have sunk to the seabed, mainly near the coast.

These findings have two important consequences. *First*, they mean that the sins of the past can be at least partially made up for by systematically cleaning the affected beaches and coastal waters of the larger pieces of plastic and then properly disposing of the (old) plastic waste collected in this way.

Second, they mean that the countries causing the plastic problem are themselves suffering most of the consequences of their waste management deficits. It is they themselves who have to cope with blighted beaches, polluted coastal waters (including consequences for tourism), and the degradation of a whole range of local wildlife species.

[4] Lebroton et al. (2019b), Chassignet et al. (2021), Wu et al. (2022), OECD (2022)

Contrary to what some reports suggest, plastic waste exports from industrialized countries only play a minor role here. Until 2017, almost all plastic waste exports from industrialized countries went to China, where they were treated, albeit at a lower quality. In any case, even at its peak, this was only less than 5% of global plastic waste. In recent years, due to new import regulations imposed by China (and increasingly other Asian countries as well), global plastic exports have declined dramatically, i.e., plastic waste is now largely disposed of domestically (or within the EU, resp.).[5]

The Facts – The West

Per capita plastic consumption in the West (USA + EU) has been about three times higher than the world average for decades, see Fig. 14.4.

However, almost 100% of the resulting plastic waste is subject to modern and proper waste treatment. An exception is waste that does not end up in household waste but is carelessly discarded/left in the environment; this is given a flat rate of 2% of plastic waste in most studies.

The overall picture regarding the treatment of plastic waste is shown in Fig. 14.5, which indicates that recycling of plastics still plays only a minor role even in the West.

What is clear, however, is that the West's contribution to the problem of marine plastic litter is very small, in the range of a few percentage points, despite its high plastic consumption.

[5] In addition, a legally binding agreement was signed in 2019 by almost every country in the world (part of the Basel Convention) that imposes much stricter rules on the export of plastic waste.

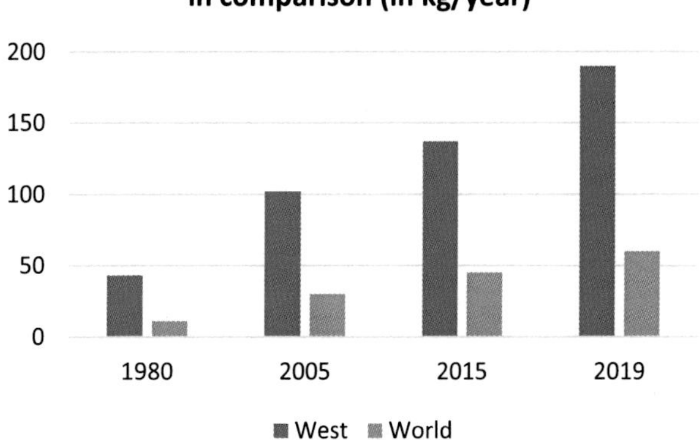

Fig. 14.4 Plastic consumption per capita in the West (USA + EU) and in the world, 1980–2015, in kg. (Source: Plastic Insight)

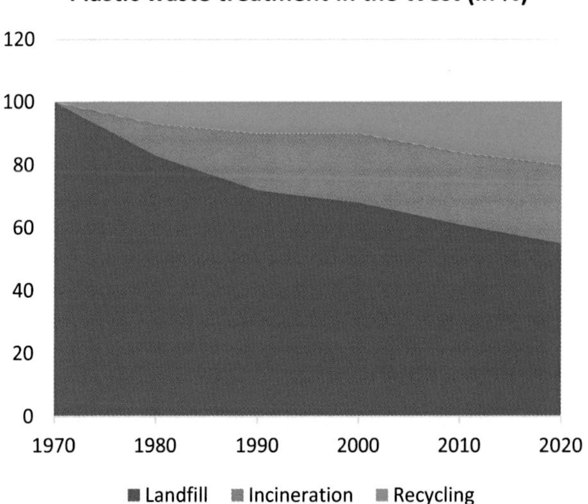

Fig. 14.5 Plastic waste by waste treatment categories in USA + EU, 1970–2020, in %. (Sources: Plastics Europe, USA EPA, own calculations)

Projection Until 2050

Figure 14.3 suggests that mismanaged plastic waste worldwide is likely to have reached a maximum of about 80–90 million tons/year a few years ago; the same is then essentially true for new plastic entering the ocean at ca. 2 million tons/year.

The same result is reached by the following reasoning.[6] There is a close correlation between the quality of waste disposal and economic development in a country (see Fig. 14.6).

Most countries in Asia will (continue to) develop rapidly – especially China and India, which were responsible for 40% of all untreated plastic waste in 2015. Therefore, a significant improvement in waste management can be expected there in the coming decades.[7] The development in Africa is unclear – there could well be an increase in problematic plastic waste here –, but overall there should be a significant reduction in the quantities of plastic newly entering the sea. This can also be expected because there is a growing awareness of the problem of plastic waste in many affected countries and because, as we have seen, measures to improve waste management have immediate and visible positive consequences for their own environmental situation.

In summary, it is realistic to expect that by 2050, quantities of plastic in the order of 30–50 million tons will once again enter the sea, i.e. mainly into the near-shore waters, especially along many of Africa's coasts.

---

[6] See also Lebroton (2019a).

[7] China has already adopted a plan of action in 2015, according to which plastic discharges into the sea should be halved by 2020. According to Bai and Lee (2020), this goal has been achieved.

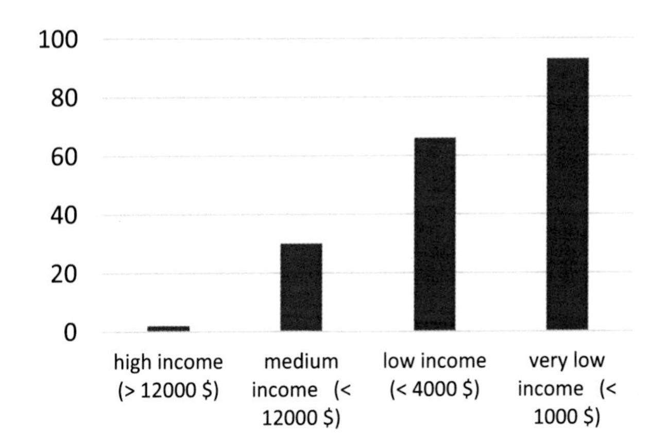

Fig. 14.6 Improperly treated waste as a % of all waste by income category (in GDP per capita, US$ by purchasing power). Data from 2015. (Source: World Bank 2019.) (Currently, there are 80 countries in the "high income" category, "medium-high income" = 60 countries, "low income" = 50 countries, "very low income" = 30 countries)

Evaluation/Summary

There is no doubt that plastic consumption will continue to increase worldwide, especially in many Asian and African countries. The advantages of plastic over possible alternatives are simply too significant for a reversal of the ongoing trend to be realistic. But regardless of whether or not plastic consumption *could be* significantly curbed, it is not at all clear to what extent or in what areas it *should be* curbed. Intuitively, it seems clear that, for example, reusable packaging is "better" than (plastic) disposable packaging;

however, detailed scientific studies sometimes come to very different results.[8]

In other words, the triumph of plastic over the last 70 years has tangible positive aspects even from a purely ecological point of view. Therefore, it cannot be a question of condemning plastic consumption across the board and pushing it back. In detail, there is certainly still a lot of progress to be made in terms of rationally trading plastic solutions off against the use of other materials, and in this way some plastic applications that are common today might disappear. On the other hand, there will also be new reasonable applications for plastics in the future.

Regarding the topic of **plastic recycling**, mankind is still rather at the beginning; and we should make considerable efforts to increase recycling rates. However, the extent to which the structural disadvantage of plastics in this area can actually be overcome by new technologies is an open question.

Ultimately, however, this is not a decisive question either. Even if a circular economy for plastics (unlike for metals) should not be achievable or should lie in the distant future: as long as plastic waste is either incinerated or properly landfilled, there is no real problem.[9]

Instead, the core problem lies in the fact that in a number of countries, especially in Asia and Africa, large amounts of plastic waste end up somewhere on the ground rather than in incinerators or properly managed landfills. And well over 90% of this is not imported plastic waste from industrialized countries, but waste produced by the country itself. To put it bluntly: Even if the West – the USA and the EU – were to completely end their plastic consumption

[8] See, for example, the study on the ecological comparison of plastic bags vs. other containers on OurWorldinData (Plastic).

[9] The fabrication and incineration of plastics does produce CO_2; at current levels this accounts for 3–4% of annual CO_2 emissions.

overnight, the problem of plastic waste in the oceans would remain virtually unchanged.

By far the most urgent task with regard to plastic is therefore neither the avoidance of plastic products nor the expansion of recycling, but the establishment of functioning waste management in the poorer countries. The industrialized countries should provide much more support in this area than they have in the past. Significant progress has already been made in recent years (especially in China, which has been the main polluter up to now), and this will mean that "horror scenarios" ("more plastic in the oceans than fish") will not come to pass.[10]

As for the ecological consequences of plastic waste, they are primarily *local* rather than *global*. In contrast to the CO_2 emissions causing climate change, the majority of the "plastic emissions" into the sea remain local, and the countries causing them (including some of the animal species living there) are themselves the main sufferers.[11]

The plastic waste that leaves coastal waters does cause the visible huge plastic accumulations in the oceans. Hopefully, these can be collected again at least partly (one can probably assume that the numerous ideas/activities in this direction will be successful in the long run); besides, these larger plastic parts should cause little lasting damage in the end despite some very negative consequences.[12]

In contrast, the repercussions of **microplastics**, which result from the slow decomposition of large plastic pieces,

[10] If there were to be 100 million tons of plastic in the ocean in 2050, that is only a small fraction of the 6–10 billion tons of fish.

[11] It is estimated that a total of about 500 animal species worldwide (especially turtles, seals, whales, seabirds) are significantly affected by marine plastic waste, see OurWorldinData (Plastic Pollution).

[12] Already many whales, other large animals, and countless seabirds have lost their lives because they mistook plastic debris in the ocean for food and swallowed it. This is bad, but on the whole these are isolated cases that do not contribute significantly to the endangerment of animal species.

are unclear and potentially more critical. These very small plastic particles are now found all over the planet – from Arctic ice to ocean floors 10 km below the surface. They are mainly ingested by marine organisms and can thus accumulate in the food chain, all the way up to humans. This ubiquitous presence of tiny plastic particles is significantly increased by microplastics from other sources (tire abrasion, washing of synthetic textiles, artificial turf, etc.).

It is true that, despite intensive research, so far there is no evidence of health problems caused by these particles, neither in humans nor in most (marine) animals; but this must be qualified as an intermediate state of scientific knowledge at the moment.

To summarize: It is not plastic consumption as such that is the problem, but the fact that modern waste management and technologies are not yet widespread, and therefore relatively large amounts of plastic waste are released untreated into the environment. The main focus of action in the next decades should consequently be,

- to drastically improve this situation,
- to retrieve as much plastic as possible from the sea and especially from coastal waters
- and to find ways to reduce microplastics from other, direct sources.

None of this will lead to the complete disappearance of plastic in the oceans. Indeed, it is probably typical of many new technologies that there are downsides as well as good sides. The rational response to this, however, is not to simply reject this technology and thus give up its positive effects on health, standard of living, educational opportunities, or whatever the case may be. Rather, it is to manage the downsides effectively, i.e., to reduce or compensate for them as far as possible.

15

Dead Zones in the Oceans – The P/N Cycle

Introduction

"Water is life" – we already learn this in elementary school. All life on our planet originated in water, and even today more than half of all higher developed animal species live wholly or partly in water. For some years now, however, there has been increasing talk of "dead zones in the sea" as one of the global ecological hotspots. What is this all about?

The term "dead zone" is indeed misleading. It refers to areas in the ocean that have oxygen levels that are well below average (<2 mg/L compared to normal levels of about 10 mg/L). Such zones also occur in nature, and a whole range of living organisms have adapted precisely to these environmental conditions. In other words, dead zones are anything but dead, but they are actually uninhabitable (i.e., deadly if they cannot escape quickly enough) for more highly evolved marine animals – primarily fish and crabs –, and that is why they bear their name.

© Springer-Verlag GmbH Germany, part of Springer Nature 2022

T. Unnerstall, *Factfulness Sustainability*,
https://doi.org/10.1007/978-3-662-65558-0_15

Thus, the problem discussed in this chapter centers around the oxygen content in seawater. It normally varies between 8 and 12 mg/L depending on temperature, water depth, salinity, and other factors. Human activities affect this oxygen content in two very different ways:

1. by emitting CO_2 in large quantities
2. by inputs of phosphorus (P) and nitrogen (N) from fertilizers, which enter the sea mainly via rivers.

Re (1)

The oceans absorb 30–40% of anthropogenic CO_2 emissions. As a result of this and climate change, oxygen concentrations in seawater are affected due to complex mechanisms. On the one hand, there is a general decrease in the oxygen content of the oceans – so far by about 2%, by 2100 by about 7% according to current predictions. On the other hand, the dead zones in the oceans have increased significantly, but they still account for far less than 0.1% of the ocean areas. I will not discuss these effects in detail below, since they are a consequence of CO_2-emissions and not an independent problem.

Re (2)

The fertilizers phosphorus (P) and nitrogen (N) applied to croplands are only partly absorbed by the plants; the rest gets into the soils, partly into the groundwater, and partly also (via various processes) into rivers and from there into the sea. Depending on local conditions, this can lead to unnatural enrichment of P and N in seawater in the nearer or wider vicinity of river mouths. Since P and N are also nutrients (i.e. "fertilizers") for aquatic plants, such nutrient enrichment – known as "eutrophication" in technical jargon – often leads to increased, unnaturally strong growth of short-lived algae near the sea surface. When these algae die and sink to the bottom, they are decomposed by certain bacteria through high oxygen consumption. This oxygen consumption can then finally lead to oxygen deficiency or even dead zones in deeper water layers in the affected marine area.

In principle, this mechanism:

leaching of P and N from croplands via rivers into the sea.

→ possibility of P/N accumulation in certain marine zones.

→ increased algal growth.

→ oxygen deficiency due to bacterial algal decomposition,

is well understood. In detail, however, the corresponding processes are extremely complex and depend on a large number of local parameters (water depth, water currents, light conditions, quantity ratio of P to N, population of so-called phytoplankton, season, and many others). Different water bodies therefore react very differently to the same P/N inputs from outside or even to the same P/N concentrations. In general, however, it seems that

- shallow waters react more sensitively than deep waters,
- P plays a greater role than N, and
- marine zones are particularly endangered if they have little water exchange with neighboring waters.[1]

In any case, it is clear that the current and future extent of the ecological problem of "dead zones in the sea" depends

[1] Therefore, the sea with the greatest risk worldwide is the Baltic Sea. As a very shallow sea (average depth only 52 m), its exchange of water with the North Sea/Atlantic – varies with the weather but – is unusually low. Therefore, there are a number of oxygen deficiency zones in the Baltic Sea independent of humans, which have, however, become significantly larger due to the intensive agriculture of the surrounding countries (especially Sweden, Denmark, Poland, Germany) and the corresponding P/N inputs (see below).

primarily on the past and future use of P fertilizers, partly also N fertilizers. Before presenting the facts about dead zones, it is therefore necessary to look at the so-called P and N cycles.

The Facts – The P Cycle

Phosphorus (P), as an element of all life, is primarily bound in large quantities in the earth's crust (several million Gt), in the earth's soil (several 100 Gt) and in the oceans (about 100 Gt). However, a comparatively very small amount of about 10 million tons (= 0.01 Gt)/year is bound in a natural cycle between these spheres: Earth's crust – Earth's soil – Lakes and rivers – Oceans – Sea floor – Earth's crust (described in a very simplified way).

Since fertilizer use began worldwide in 1950, mankind has added an *anthropogenic* P cycle to this *natural* P cycle. Currently, about 20 million tons of phosphorus per year are applied to cropland in fertilizers, but plants and soil can only absorb about half of that. Around 10 million tons/year end up in rivers and lakes and from there, for the most part, in the sea. Humans have thus roughly doubled the natural P cycle.

If the phosphorus is then transported relatively quickly from the river mouth to the open sea (through intensive water exchange, ocean currents, etc.), it cannot cause any damage: the concentration of P in seawater increases only imperceptibly, even over longer periods of time. However, if P, favored by local conditions, accumulates in nearshore water zones (or sometimes in inland lakes), this eutrophication can lead to oxygen deficiency or dead zones.

Consequently, the only way to reduce existing dead zones and mitigate the risk of new ones forming is to reduce P

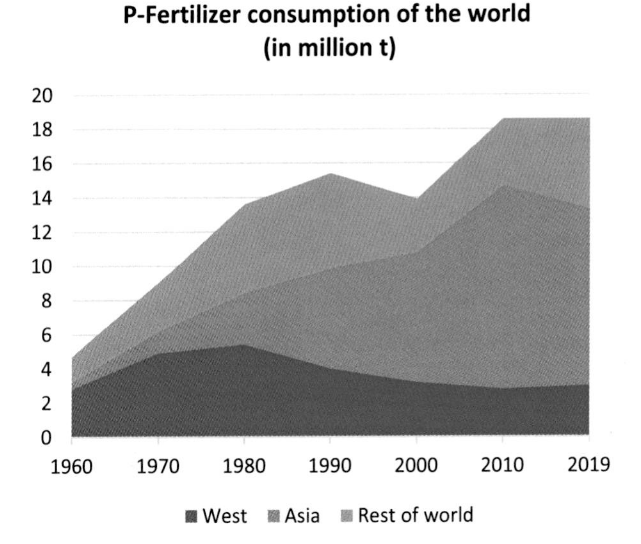

Fig. 15.1 Phosphorus fertilizer consumption in world regions, 1960–2019, in million tons. West = EU + USA. (Source: FAOSTAT)

inputs to the oceans. This, in turn, can only be achieved through more precise dosing of P fertilizers, adapted to the respective soil conditions, so that as little excess P as possible is generated and can enter the oceans.

In the West, this has largely been accomplished in recent decades.[2] Asia is rather at the beginning of this process, but at least consumption has slightly decreased there as well (Fig. 15.1). Therefore, hope is justified that the anthropogenic P cycle can be reduced again in the coming decades and thus the negative consequences in the form of dead zones can be mitigated.

[2] The often reported pollution of groundwater in Europe is not caused by P fertilizers, but by N fertilizers.

The Facts – The N Cycle

Nitrogen (N) occurs in a wide variety of chemical compounds on Earth and is involved in many processes and cycles. Mankind influences these processes in two ways in particular:

1. Similar to P, excess N fertilizers enter groundwaters ("nitrate contamination of groundwater") as well as inland waters and oceans.
2. In addition to carbon dioxide, the combustion of fossil fuels also produces various nitrogen oxides, which lead to air quality problems (particulate matter pollution, acid rain, etc., cf. Chap. 16) and also contribute to climate change.

Regarding (1.), anthropogenic N inputs to the sea are currently estimated at 40–50 million tons/year. They are thus significantly higher than the natural N input of 10–20 million tons/year. Similar to P, these inputs to the oceans do not matter *globally*, given the tremendous amounts of N bound in seawater (≈1000 Gt). *Locally*, however, under certain conditions they can contribute to eutrophication and thus to the formation of oxygen deficient zones in the vicinity of river mouths.

Against this background, looking at N fertilizer consumption worldwide yields Fig. 15.2. In the West, consumption has stagnated with a slight downward trend since 1980, and in Asia the steady increase over decades seems to have come to a halt. The expectation is therefore that the considerable N surpluses – of the 110 million tons of N fertilizers applied to arable land, about 40% ultimately end up in the sea – can be significantly reduced over the course of the century by the spread and further development of modern agricultural techniques.

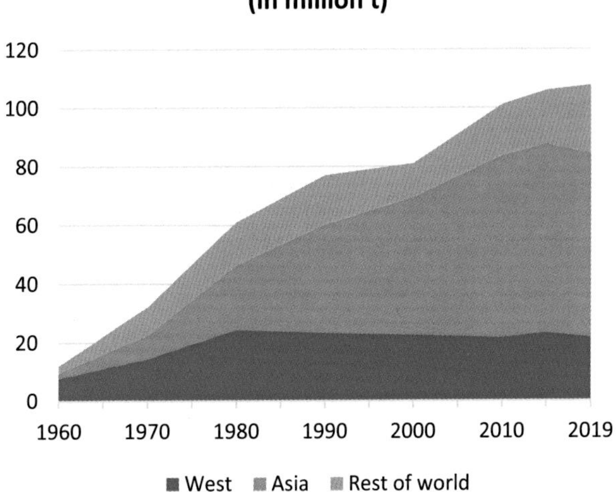

Fig. 15.2 Nitrogen fertilizer consumption in world regions, 1960–2019, in million tons. West = EU + USA (Source: FAOSTAT)

The Facts – Dead Zones in the Oceans[3]

Dead zones created by P/N inputs can occur around estuaries on the continental shelf of the oceans, as described. These so-called "shelf seas" extend on average to about 70 km offshore, are generally up to 200 m deep, and account for about 30 million km^2 worldwide, or ca. 8% of the Earth's total ocean area.

As far as is known, dead zones currently cover an area of about 0.25 million km^2 (<1% of the shelf seas). They are strongly concentrated in five regions:

[3] As outlined in the introduction, I discuss here only the dead zones created by eutrophication (i.e., anthropogenic P/N inputs to the ocean) and thus nearshore, not the dead zones in the open ocean caused by climate change.

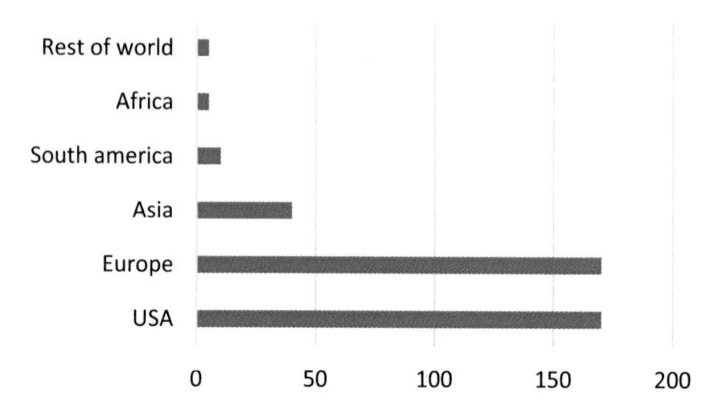

Fig. 15.3 Number of dead zones by world region, 2008 data (more recent data are not available). (Source: Diaz and Rosenberg 2008)

- Baltic Sea – 80,000 km^2 – many rivers as sources
- Black Sea – 50,000 km^2 – main sources: Don, Danube
- North Sea – 35,000 km^2 – many rivers as sources
- Gulf of Mexico – 22,000 km^2 – main source: Mississippi
- East China Sea – 15,000 km^2 – main source: Yangtze River

Beyond this total of 0.2 million km^2, there are about 400 known dead zones of generally much smaller dimensions. Their geographical distribution is shown in Fig. 15.3.

Europe and the USA are by far the most affected world regions. This may be because high fertilizer use has been practiced here the longest. However, one must also consider the possibility that in the West coastal waters are under much closer monitoring, i.e. that smaller dead zones may not yet have been discovered in other regions of the world.

In Europe and the U.S., the phenomenon of eutrophication and its consequences has long been the subject of political attention; thus, targeted programs to reduce fertilizer inputs to the oceans (and also to improve wastewater treatment) have been in place for decades. This has been successful in most cases, even though it seems to take decades before the affected marine zones begin to normalize. Nevertheless, there have already been measurable improvements in the North Sea, the Black Sea and off the east coast of the USA.[4] In the Gulf of Mexico and the Baltic Sea, they are expected.[5]

The main focus for the future is therefore on developments in other regions of the world, especially in Asia.

Here, the discovery of new dead zones has no longer increased (cf. Figure 15.4). The further development is open; but today, the problem is still manageable:

- a maximum of 0.2% of the shelf seas are affected,
- in the affected zones, oxygen deficiency does not occur permanently, but only seasonally or only from time to time.

The most critical areas are the Chinese and Japanese coastal waters.

Evaluation/Summary

The (P/N-caused) dead zones in the oceans are essentially *not* a global ecological problem. First, on a global scale, the underlying anthropogenic P/N inputs to the world's oceans

---

[4] Where P/N inputs have been drastically reduced, it looks as though the affected ecosystem does not completely re-evolve to its original condition, but the number of animal species increases significantly again.

[5] In the Baltic Sea, an improvement of the situation is complicated by the fact that, due to climate change, the exchange of seawater with the North Sea tends to decrease even further.

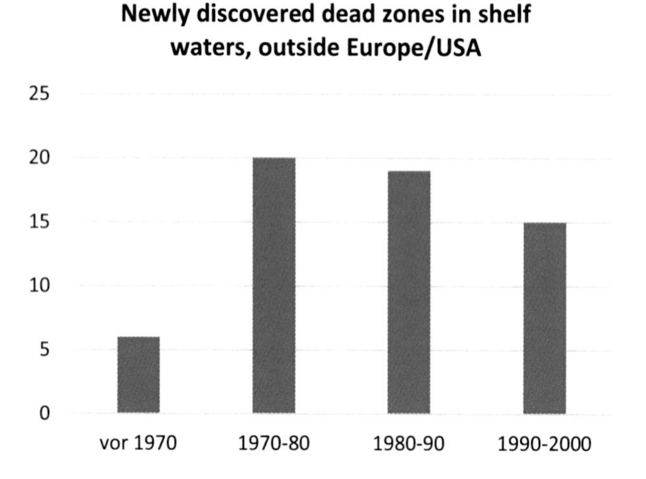

Fig. 15.4 Time course of discovery of dead zones outside USA and Europe, until 2000 (more recent data not available). (Source: Diaz and Rosenberg 2008)

are negligeable for the foreseeable future, having only an imperceptible effect on the P/N content of the oceans. Second, while such dead zones do exist around the globe, in total they represent less than 1% of coastal waters (and less than 0.1% of the oceans as a whole). Third, and most importantly, the causal chains that lead to them are essentially *local:* dead zones off a country's coast are caused almost exclusively by that country's agricultural practices.[6]

Moreover, in the end the ecological consequences of even larger dead zones are manageable. Vertebrates are usually able to escape,[7] and the more severely affected species are generally not threatened in their populations. Finally, it should be noted that the problem in the most affected

[6] An exception is the Baltic Sea, whose oxygen content depends on the P/N inputs of several riparian states.

[7] This effect entails that not infrequently local fisheries even tend to benefit because they encounter untypically high stocks of fish in the vicinity of dead zones.

regions – USA, Europe, meanwhile also China – is the subject of long-term, scientifically accompanied improvement measures and will therefore with some certainty not spread/get worse here.

It is certainly too early to give the all-clear. But on the priority list of global joint action, other issues are clearly higher up (cf. Chap. 2).

Excursion: Dead Zones in Lakes

Fertilizer surpluses from agriculture in the form of phosphorus (P) and nitrogen (N) enter not only the oceans but also inland waters via rivers. Under appropriate local conditions – shallow water depth, little water exchange – the phenomenon of eutrophication can also occur in lakes and, in some cases, result in oxygen deficiency zones.

For about 20 years, the 1000 largest lakes in the world[8] have therefore been monitored by satellite for their eutrophication status. The results of this monitoring are comparable to those for coastal waters/shelf seas: Less than 1% of lakes have critically high nutrient concentrations, i.e., oxygen deficient areas. Elevated nutrient levels – but to a degree that is not uncommon in nature – are present in another 10–15% of lakes.[9]

Similar results have been obtained by scientific research programs that are monitoring the ecological condition of the approximately 100 largest lakes in the world in greater detail. About 15% of these lakes show major changes in their ecological status (i.e., especially nutrient

[8] An exception is the largest inland water body on earth, the Caspian Sea, which plays a special role as a saltwater lake.

[9] Overall, the greatest problems are found in artificial reservoirs, which are extremely shallow (average depth 16 m) and therefore particularly sensitive to P/N inputs.

concentrations); in a few with less than 1% of the water volume, there are critical P/N values and dead zones (see Fig. 15.5).

It is noteworthy that since the 1970s, the condition of lakes in the West and in the northern regions of the world has improved noticeably on average, while in the tropics it has tended to deteriorate during the last decades.[10]

Even though it takes decades for policy measures to reduce P/N inputs to have an impact, in the long term it is possible to at least bring the ecological status of lakes back closer to natural conditions.

Excursion: Planetary Boundaries – P/N Cycle

The concept of "planetary boundaries" was developed in Stockholm about 10 years ago and has since received considerable international attention. Similar to the "ecological footprint," it is intended to provide a comprehensive assessment of the extent to which human activities threaten the stability of ecosystems and the future livelihood of humankind.

The concept focuses on the following ecological hotspots (simplified):

- (Fresh) water consumption
- Species extinction
- Deforestation ("Land System Change")
- P-cycle and N-cycle
- Air pollution
- Introduction of novel substances/organisms into the environment

---

[10] By far the largest lake with critical P/N values and oxygen deficiency zones is Lake Victoria in Africa. Recent studies (Myanza et al. 2018) show that the situation here has at least not deteriorated further since about 10 years ago.

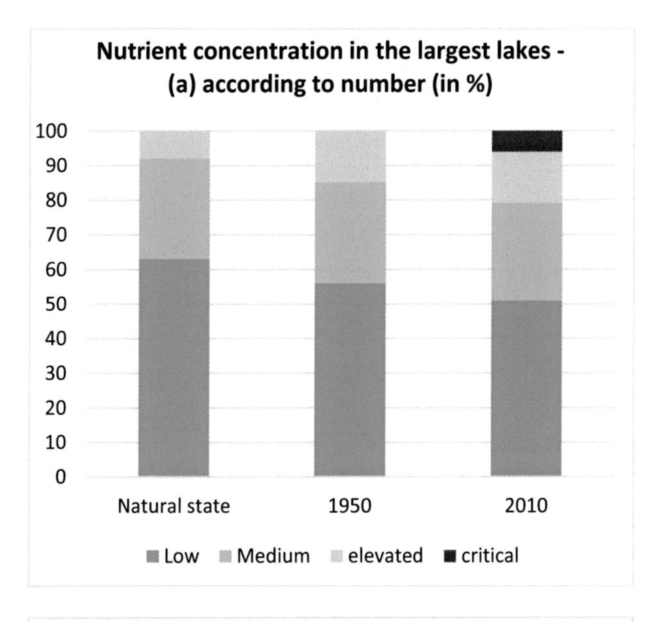

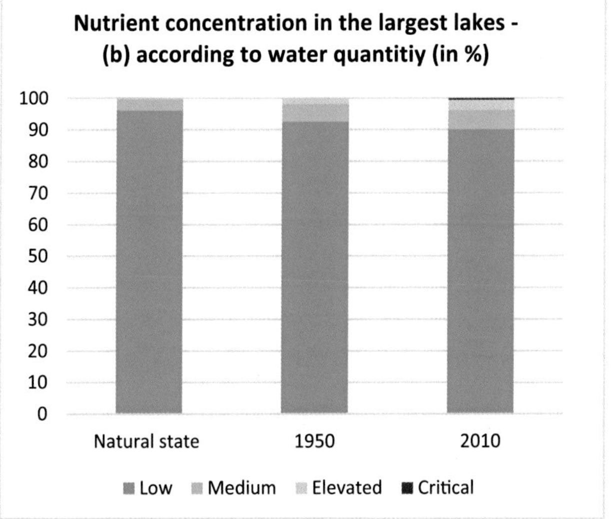

Fig. 15.5 Nutrient concentrations in the largest ca. 100 lakes on Earth in four categories, by 2010, in % **(a)** measured by number, **(b)** measured by water volume. (Sources: Izmailova and Rumyantsev 2016, own calculations)

- Ozone hole
- Climate change/ocean acidification

I do not go into the details of this concept in this book, since a separate chapter is devoted to the most important topics – except for climate change – anyway. But if you are familiar with this concept and the corresponding illustrations in the media, you may remember that it is precisely the P/N cycle that is classified as particularly critical there. A "high risk" is asserted that the current state of the P/N-system will exceed the Earth's stress limits. How does that fit with the assessment in this book?

Well, first, there are still serious scientific differences of opinion on the details of the relevant mechanisms and quantities of the P/N cycle. Second, the "high risk" classification is based on assumptions and extrapolations that must be considered highly uncertain. Other scientists therefore consider the planetary boundaries with respect to the P- and N-cycles as still preserved, thus reject the "high risk"-classification of the Stockholm researchers.

Third, the concept of planetary boundaries does not make any statement about how serious the various ecological problems are to be rated in *comparison with each other*. Indeed, the data in the present chapter show that the P/N cycle is neither a truly global problem in the strict sense, nor are the present and foreseeable impacts nearly as severe as, say, forest loss (or of course climate change), simply in terms of the size of the ecosystems affected.[11]

[11] In my view, the overall concept of "planetary boundaries" suffers from similar methodological difficulties as the concept "ecological footprint." Presenting complex, completely different problem areas on a single, uniform scale has the advantage of being highly descriptive and enables catchy statements. But the price is high: the risk of not really doing justice to any of the topics in the end and tending to distort the view of the respective real situation.

16

Pollutants in the Environment

Introduction

December 8, 1970, is a historic date for environmental protection: The world's first environment ministry was founded in – who would have thought it? – Bavaria. Today, more than 50 years later, the Bavarian state government's justification at the time and the German government's first environmental programs adopted in the years that followed, seem quite surprising. The ecological hotspots of today, which I have examined so far in this Part IV, play practically no role. The focus is rather on a completely different topic: keeping the air and water clean.

This becomes understandable, however, if one considers the environmental situation in Germany at the time. The air, especially in the big cities, often had high concentrations of pollutants: sulfur dioxide (SO_2), nitrogen oxides (NO_x), carbon monoxide (CO), lead, VOCs, etc. The Rhine river was a cesspool, full of toxins from industry and household sewage, with extremely low oxygen levels over long stretches; biologically, it was as good as dead. In other words, the environmental problems at that time were not of a long-term, insidious, indirect nature – as is the case today with climate change, species extinction, rainforest deforestation, and resource consumption – but were acute, visible

© Springer-Verlag GmbH Germany, part of Springer Nature 2022

T. Unnerstall, *Factfulness Sustainability*,
https://doi.org/10.1007/978-3-662-65558-0_16

to everyone, and had immediate adverse effects on the health of the population.

In the following decades, therefore, comprehensive air and water pollution control measures were implemented in Germany, in many countries in Europe and in the USA: Flue gas desulfurization and denitrification for power plants, catalytic converters for cars, development of modern sewage plant technology, strict regulations for industrial wastewater, and many more. As a result, air and water quality with regard to pollutants have steadily improved in most industrialized countries and are now generally at a very good level.Swimming in the Rhine is allowed again, the last "smog alarm" dates back to 1991, and "acid rain" is also history.

A similar development can be observed worldwide in recent times – at least with regard to air quality[1] . Emissions of SO_2, CO and NO_x peaked globally 5–10 years ago. Air quality problems in this respect still occur mainly in Asia (China, India, partly in the Middle East)[2] but are increasingly more local and temporary in nature.

Moreover, this problem will improve significantly until 2050 in most countries anyway. The main cause of these pollutants is the combustion of fossil fuels in car engines and power plants; and these will (have to) be successively replaced by alternative motors and regenerative or nuclear energies in the next decades, due to climate protection.

---

[1] With regard to water quality, comprehensive data are available from too few countries to be able to adequately assess the situation or the development over time globally.

[2] With regard to air quality, there is now a fairly tight network of measuring stations worldwide (with the exception of Africa) (WAQI.info); as a rule, the values for particulate matter (PM_{10}, $PM_{2.5}$), ozone, NO_x, CO and SO_2 are measured continuously.

However, there are two pollutants in particular that continue to have global significance and are accordingly the subject of intensive observation, research and political measures worldwide: **Particulate matter** and **mercury**. I will therefore confine myself in the following to these two issues.

The Facts – Particulate Matter

A glance at the online WAQI.info world map of air quality quickly reveals that exposure to particulate matter (PM_{10} and $PM_{2.5}$) is by far the most serious air pollution problem, uniformly on all continents. So-called $PM_{2.5}$ is particularly critical: these are dust particles which, due to their small diameter of less than 2.5 μm – unlike the coarser dust PM_{10} –, are generally not intercepted by the protective mechanisms of the human body. In other words, they can enter the human lungs and cause long-term damage to health. It is therefore a serious environmental issue, though actually not an *ecological* one. The planet's ecosystems are hardly affected by particulate matter; higher concentrations are mostly confined to cities and their immediate surroundings.

This observation already points to the main sources of particulate matter: The combustion of coal, gasoline, and fuel oil is responsible for about half of $PM_{2.5}$; the combustion of wood, in Europe primarily in fireplaces, also plays a significant role. There are mechanical sources (such as tire abrasion in traffic); and, finally, natural sources (dust from dry soils/deserts, sea salt) are also involved to an average of ca. 20% – obviously varying greatly from region to region.

Let us look at the temporal development of particulate matter pollution over the last decades in Fig. 16.1:

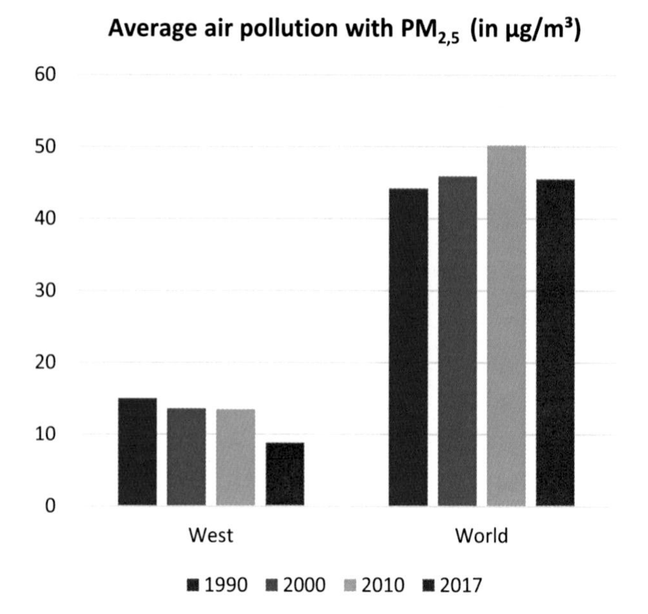

Fig. 16.1 Average $PM_{2.5}$ exposure for the population, 1990–2017, in µg/m^3, West = USA + EU. (Sources: State of Global Air, OurWorldinData (Outdoor air pollution))

In the West, there has been a steady improvement in average pollution since 1990, and in the world as a whole, levels are also slowly declining after a peak around 2010. The highest concentrations are in southern Asia (India = 91 µg/m^3), central Africa (Nigeria = 72 µg/m^3), and the Middle East (Saudi Arabia = 88 µg/m^3). In Asia and Africa, the main cause is often domestic wood burning,[3] in the Middle East natural sources from nearby deserts play a major role.

[3] Indoor air pollution is therefore usually the much more serious problem in these countries.

Perhaps even more relevant is another way of looking at it: namely, what percentage of the population in a country, region, or the world is exposed to an average $PM_{2.5}$ concentration that is above the limit considered to be safe.

Now the problem is that just this limit is scientifically difficult to determine and therefore no uniform limit exists. The WHO sees it at 10 μg/m³, the EU has currently set it at 25 μg/m³. Figure 16.2 shows that these two limit values are worlds apart: In the West, while 50% of the population is now exposed to levels higher than 10 μg/m³, only 2% are exposed to the higher level of 25 μg/m³. Worldwide, the figure is 91% versus 67%.

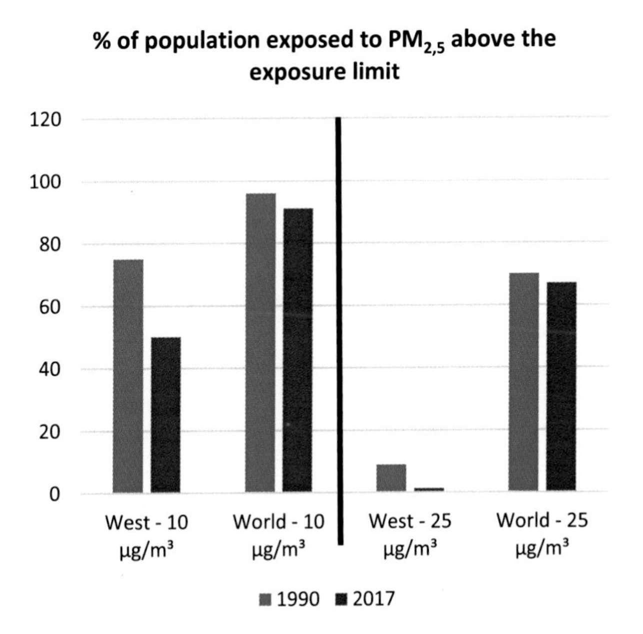

Fig. 16.2 Percentage of population exposed to average $PM_{2.5}$ above 10 μg/m³ and 25 μg/m³ for 1990 and 2017, in the West (EU + US) and the world. (Source: UNEP 2019)

Evaluation/Projection – Particulate Matter

The **conclusion** from the available global data regarding particulate matter is this. On the one hand, there has been progress in the last decades, on the other hand – especially in South Asia and Africa – there is still a long way to go to reach a satisfactory situation.

However, the **forecast** for the next decades is clearly positive. Two considerations are crucial:

- The fight against climate change – regardless of the degree to which it is successful (even in the literal sense) – will lead to a drastic reduction of the burning of fossil energy sources and thus of the main cause of $PM_{2.5}$ pollution by 2050.
- Moreover, in the course of economic development, at least in Asia, it can be expected that the use of wood as a major source of energy in households will be gradually replaced by more modern forms of energy.

We can therefore rather safely assume that, by 2050, today's air quality problems will be as forgotten in most countries as the smog alarms of the 1970s or the strict ban on swimming in the Rhine are now forgotten in Germany.

The Facts – Mercury

Throughout history, and especially in the last century, human activities have released many pollutants into the environment that are toxic to humans and animals: Pesticides,

POPs[4] (dioxins, PCBs), lead and other heavy metals. Since these are persistent substances that can be transported long distances through the atmosphere and the planetary water systems, they are now found around the globe even in the most remote areas. Due to appropriate global measures in recent decades – banning leaded gasoline, banning PCBs, etc.[5] – the concentrations have mostly decreased significantly.

Mercury is probably the most important of these pollutants and toxins. This is mainly due to two special features:

- Mercury – unlike the other pollutants mentioned above – was already used on a significant scale in the nineteenth century
- Mercury also naturally runs in a continuous cycle between land, oceans, and atmosphere, and in this way can circulate for decades: Sins of the past have a particularly long-lasting effect here.

Let us therefore take a closer look at this problem.

Mercury (Hg) is found in large quantities in the earth's crust (≈100 Gt). Humans have used only a tiny fraction of this (about 1 million tons = 0.001 Gt), but about 0.4 million tons have been emitted into the atmosphere in this way, especially in the nineteenth and twentieth centuries. The time course of these emissions is shown in Fig. 16.3.

The *natural* cycle between land, oceans, and atmosphere in the order of 2000 tons/year is overlaid by an

[4] POP = Persistent Organic Pollutants. They include the particularly toxic polychlorinated biphenyls (PCBs), which have been banned worldwide for 20 years.

[5] With regard to the particularly toxic substances, global cooperation has worked quite well in recent decades – especially when an immediate health hazard to humans could be clearly demonstrated. The best examples are the Montreal Convention (1987) on CFCs (which used to cause the hole in the ozone layer), the Stockholm Convention (2001) on POPs, and the Minamata Convention on Mercury, 2013 (see Evaluation/Projection).

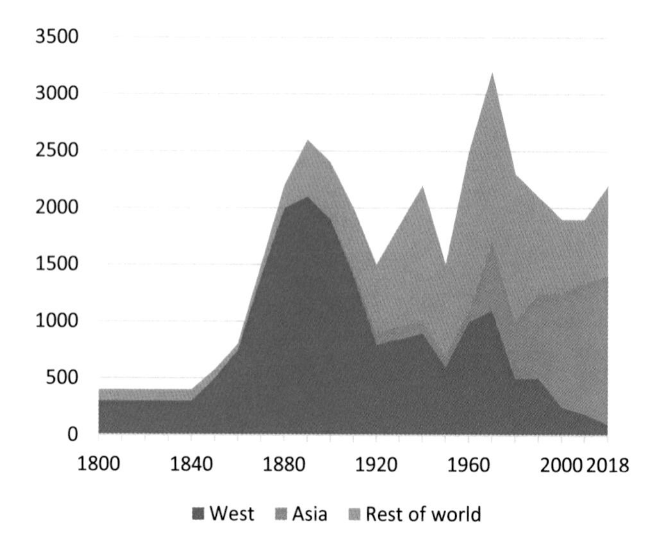

Fig. 16.3 Mercury emissions to the atmosphere, 1800–2018. (Source: Chen et al. 2018)

anthropogenic cycle of currently 3000–4000 tons/year, in which historically emitted quantities of mercury circulate for many decades.[6] The quantities leaving the cycle are deposited in the ground or on the seabed. Finally, new Hg emissions are added each year, currently on the order of 2000 tons/year. Thus, the global mercury cycle caused by humans has a total volume about 2–3 times higher than the natural cycle.

Where did mercury come from and where does it come from? Figure 16.4 shows the major **sources** today.

[6] This means that even if mercury emissions were completely stopped today, 100 years from now Hg concentrations in the atmosphere and oceans would still be significantly elevated above natural levels.

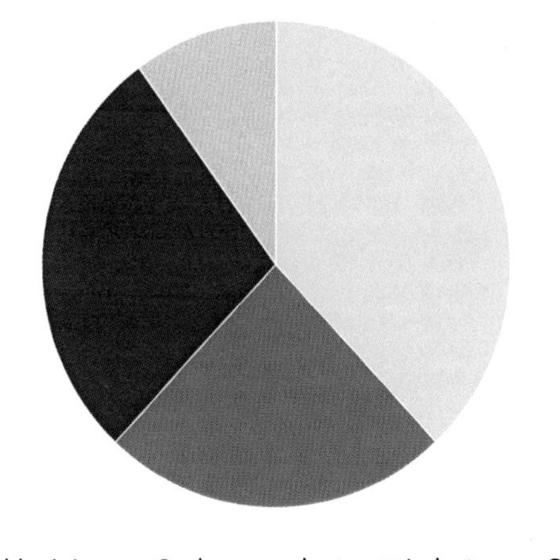

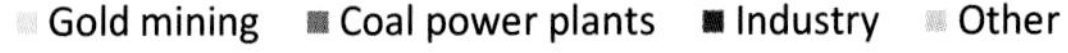

Fig. 16.4 Major sources of mercury emissions to the atmosphere in %, 2015 data. (Source: UNEP 2018)

The most important factor is gold mining – not mining on a large industrial scale[7] but small-scale, largely manual gold mining by small companies or family-owned businesses. Here, the so-called mercury process[8] is still being used today – as it was during the "gold rush" in the second half of the nineteenth century (especially in the USA) which caused the first peak in Hg emissions visible in Fig. 16.3. The resulting mercury vapors could actually be captured by recovery devices; instead, they often simply enter the atmosphere and become part of the global Hg cycle.

[7] On an industrial scale, the cyanide leaching process is used, which also poses significant – but mainly local – environmental hazards if not handled properly.

[8] Gold-bearing rock is first sieved in water until the gold concentration is as high as possible. This gold-rock slurry is then mixed with mercury, which forms a liquid alloy (amalgam) with the gold. The residual rock is screened out, the alloy is heated, the mercury evaporates, and what remains is pure gold.

The second largest single factor is coal-fired power plants around the world, which still account for a quarter of current Hg emissions. The remaining emissions are caused by various industries and mercury-containing products in waste.

The **effects** of the substantial additional mercury circulating are measurable everywhere on Earth: in the atmosphere, in the Arctic ice, in surface waters, in the depths of the oceans, in the soil. However, the main problem in terms of concrete ecological impact is not the occurrence in the environment, but the accumulation of mercury in the marine food chain and subsequently in the human body.

For this reason, mercury concentrations in fish, whales, dolphins, seabirds, polar bears, etc., and their effects have been the subject of scientific research ever since around 1970. In humans, especially blood and breast milk have long been studied for mercury levels.

With few exceptions, the available long-term studies show a consistent trend. Since the 1970s, the relevant mercury levels have been decreasing, both in animals and in humans. In some fish species[9] (tuna, sharks) and some cetacean species, values are still measured at which health hazards cannot be ruled out. In humans, on the contrary, almost everywhere[10] Hg-concentrations have reached a level that can be considered harmless (cf. Fig. 16.5).

[9] Cf. UNEP 2018. For this reason, there is a dietary recommendation, especially for pregnant women, to avoid these fish species.

[10] The values given in the figure are an average across all continents and population groups. There are, of course, differences: in the southern hemisphere, the values are generally somewhat higher than in the northern hemisphere. However, the limit values are largely complied with everywhere, with one notable exception: studies on African children show high values in some cases, which can probably be explained, among other causes, by the fact that many children there have to work in or live close to gold mines and are therefore directly exposed to Hg emissions.

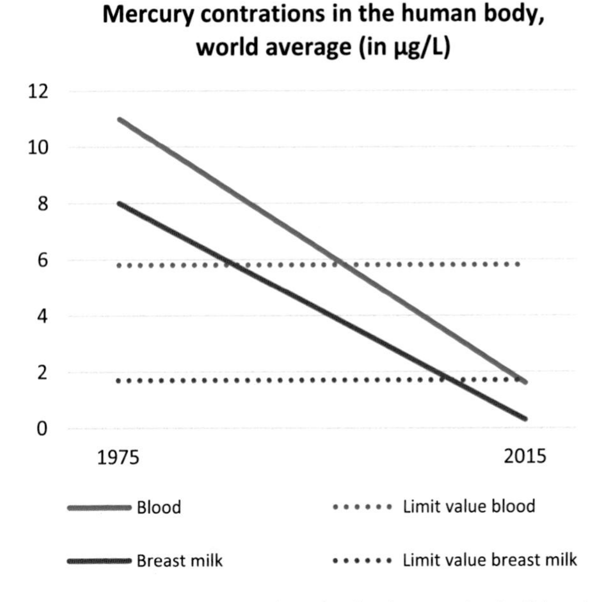

Fig. 16.5 Mercury concentrations in the human body (blood and breast milk) compared to the respective limit values, world average, 1975–2015. (Source: Sharma et al. 2019)

Evaluation/Projection – Mercury

The use of mercury by humans has a long and eventful history: from its use as a remedy as early as 2000 years ago, to its paramount role for gold prospectors on the American West Coast in the nineteenth century, to a variety of mercury-containing products (lamps, dental fillings, thermometers, weedkillers, and many others) in the twentieth century. Then, in the 1960s and 1970s, the harmful effects of mercury on humans and animals became increasingly clear. Consequently, mercury concentrations in the environment and the mechanisms of global dispersal advanced to the subject of research, and the reduction of mercury pollution became a major motive of environmental policy.

Today, it can be summarized that mankind is on the right track: emissions have been drastically reduced, Hg concentrations in the environment and in the human body are decreasing almost everywhere on earth.[11]

From today's perspective, three developments will be decisive for the foreseeable future up to around 2050. *First*, the days of coal-fired power generation as a major source of emissions are numbered for climate protection reasons. *Second*, the so-called Minamata Convention came into force in 2017, an international agreement to reduce mercury consumption in products and to mitigate mercury emissions, especially in gold mining. *Third*, it will be important – independently of this convention – how quickly this by far most important emission factor can be reduced, i.e. how fast small-scale gold mining in South America, Africa and South Asia can be turned into gold mining operating with modern protective mechanisms for the environment and the people working there.

Overall, there is a high probability that by 2050 (new) mercury emissions will have fallen significantly. This should permanently banish the dangers to human health globally.[12] But in individual marine ecosystems and animal species, the traces of the past will probably still be clearly measurable at the end of the century. Unlike the current problems around particulate matter, mercury will not be quickly forgotten.

[11] A very similar development can be asserted with respect to POPs, especially PCBs: initially, until about 1970, a sharp increase in use with corresponding emissions; then – as scientific evidence of problematic ecological and health effects accumulated – within about 20 years, a drastic reduction in production and emission levels; subsequently, with a time lag of several decades, a decreasing concentration of the pollutant in the environment and finally also in the human body. Here as well, today the measured levels in the blood are below the limit values practically everywhere in the world.

[12] On the other hand, critical pressures are likely to continue at the local level in individual cases, see footnote 10.

Epilogue

Having read this book, you might ask yourself two questions.

The probably most obvious question is this. Again and again, the reality behind the ecological headlines of our time turned out to be quite different from the picture conveyed by most media. How can this be? How is this persistent and profound gap possible, all the more since we are supposed to be better informed than ever?

Well, it turns out that this phenomenon is by no means limited to ecological issues. In fact, it holds true for just about any global issue. Fortunately, there is a very good book that analyses the main mechanisms behind this difference between the world as it is and the world as we see it: "Factfulness" (2018) by Hans Rosling, a Swedish scientist. This book – an international bestseller – gives a comprehensive answer to this first question, and I can strongly recommend it.

The second question that suggests itself in this context is: What about climate change? Could it be that the

© Springer-Verlag GmbH Germany, part of Springer Nature 2022
T. Unnerstall, *Factfulness Sustainability*,
https://doi.org/10.1007/978-3-662-65558-0

ever-growing bad news on this front – floodings in Pakistan, droughts in Europe and in Africa, melting of the Arctic ice faster than climate models predicted, record temperatures around the globe, new coal plants in China, etc. – convey a distorted picture of the real state of affairs here as well?

I do not have an answer to this question, and I doubt that it can be answered in the same straightforward, fact-based manner as I have employed in this book with respect to the other sustainability issues. There are, however, a few things that I am convinced of:

- The risk that climate change might have grave consequences even earlier and is progressing even faster than expected is too high to take it – i.e., humanity should really focus its sustainability efforts on this end and accept possibly slower progress on other ecological issues.
- The CO2 reduction path to 1.5° is not achievable anymore, but we still have a very realistic chance of keeping temperature rise in the range of 2–2.5° until 2100.
- It is wholly unjustified and counterproductive to talk about an "end of humanity" or a "suicide" in this context. We have to be constructive, innovative, positive in our outlook in order to succeed.

To end on a quite personal note… I have no idea what the main issues for humanity will be in the year 2100; but I am very optimistic that it is neither climate change nor any of the other problems discussed in this book: they will have been mastered by the next generations.

Appendix

Abbreviations

BfG	Federal Institute of Hydrology
GDP	Gross Domestic Product
CGLS	Copernicus Global Land Service
CO_2	Carbon dioxide
CH_4	Methane
DENA	German Energy Agency
EEA	European Environment Agency (Engl. European Environment Agency)
EU	European Union (28 states)
ESTAT	Statistical Office of the European Union
FAO	Food and Agriculture Organization of the United Nations
FAOSTAT	Food and Agriculture Organization Statistics
CFC	Chlorofluorocarbons
FR	Fertility rate
GFED	Global Fire Emission Database
GFW	Global Forest Watch
GHA	Global Hectars (measure)
Gt	Gigatons (= billion tons)
IAI	International Aluminium Institute

(continued)

© Springer-Verlag GmbH Germany, part of Springer Nature 2022

T. Unnerstall, *Factfulness Sustainability*,
https://doi.org/10.1007/978-3-662-65558-0

(continued)

IEA	International Energy Agency (IEA)
IPBES	Intergovernmental Platform on Biodiversity and Ecosystem Services
INPE	Instituto Nacional de Pesquisas Espaciais (Brazil)
IUCN	International Union for Conservation of Nature
LPI	Living Planet Index
LULUCF	Land Use, Land-Use Change and Forestry
NGO	Non-governmental organization
NO_x	Nitrogen oxide
OHI	Ocean Health Index
OSPAR	Oslo-Paris Convention of 1992
PCB	Polychlorinated biphenyls
POP	Persistent Organic Pollutant (POP)
PtX	Power-to-X technology
PV	Photovoltaics
RCP	Representative Concentration Pathway
SDG	Sustainable Development Goal (SDG)
t/ha	Tons per hectare
UN	United Nations (UN)
UNEP	United Nations Environment Programme (UNEP)
UNESCO	United Nations Educational, Scientific and Cultural Organization
USGS	US Geological Society
VOC	Volatile Organic Compounds (VOC)
WHO	World Health Organization
WWF	World Wildlife Fund

References

Chapter 1 – Introduction

Databases

Global Footprint Network: data.footprintnetwork.org

Books and articles

Randers J (2012a), “2052 – Der neue Bericht an den Club of Rome,” oekom, München
Schneidewind U (2018a), “Die große Transformation,” Fischer, Frankfurt

© Springer-Verlag GmbH Germany, part of Springer Nature 2022

T. Unnerstall, *Factfulness Sustainability*,
https://doi.org/10.1007/978-3-662-65558-0

Chapter 3 – "The Western economic system and way of life are unsustainable" – False!

Books and articles

Meinert S (2018), "Nachhaltiger Konsum 2030." Werkstattbericht für das BMJV (Bundesministerium für Justiz und Verbraucherschutz), bmjv.de/SharedDocs/Downloads/DE/Verbraucherportal/Nachhaltigkeit/Werkstattbericht-nachhaltiger-Konsum-2030.pdf?__blob=publicationFile&v=1
Schneidewind U (2018b), "Die große Transformation," Fischer, Frankfurt
Steffens D und Habekuss F (2020), "Über Leben," Penguin Verlag, München

Chapter 4 – The fight was worth it

Books and Articles

Meadows D (1972a), "Die Grenzen des Wachstums," DVA, Stuttgart

Chapter 5 – World population

Databases

ESTAT: ec.europe.eu/eurostat/de/data/database
UN World Population Prospects: population.un.org/wpp/DataQuery
US Census Bureau: www.census.gov/data/datasets/2017/demo/popproj/2017-popproj
World Bank: data.worldbank.org
Worldometer: worldometers.info/population/

Chapter 6 – Land use

Databases

FAOSTAT: fao.org/faostat/en/#data
Global Footprint Network: data.footprintnetwork.org
OurWorldinData: ourworldindata.org

Reports from international organizations

EU (2018a), World Atlas of Desertification: wad.jrc.ec.europa.eu/download

FAO (2011), The State of the World's Land and Water Resources for Food and Agriculture: fao.org/3/i1688e/i1688e.pdf

FAO (2012a), World Agriculture towards 2030/2050: fao.org/3/ap106e/ap106e.pdf

Books and articles

Eitelberg D, van Vliet J, Verburg P (2014), "A review of global potentially available cropland estimates and their consequences for model-based assessments," Global Change Biology 21 (3)

Molotoks A, Stehfest E, Doelman J, Albanito F, Fitton N, Dawson T, Smith P (2018), "Global Projections of future cropland expansion to 2050," Global Change Biology 24 (12)

Ramankutty N, Evan A, Monfreda C, Foley. (2008), "Geographic distribution of global agricultural lands in the year 2000," Global Biogeochemical Cycles 22

Zabel F, Putzenlechner B, Mauser W (2014), "Global Agricultural Land Resources – A High Resolution Suitability Evaluation and Its Perspectives until 2100 under Climate Change Conditions," PLOS ONE 9(12): e114980

Chapter 7 – Food

Databases

FAOSTAT: fao.org/faostat/en/#data

Reports from international organizations

FAO (2012b), World Agriculture towards 2030/2050: fao.org/3/ap106e/ap106e.pdf

FAO (2015a), Status of the World's Soil Resources: fao.org/3/i5199e/I5199E.pdf

FAO (2017a), The Future of Food and Agriculture – Alternative Pathways to 2050: fao.org/3/I8429EN/i8429en.pdf

FAO (2017b), The Future of Food and Agriculture – Trends and Challenges: fao.org/3/a-i6583e.pdf

Heinrich-Böll-Stiftung (2018), Fleischatlas: boell.de/sites/default/files/2019-10/fleischatlas_2018_V.pdf?dimension1=ds_fleischatlas_2018

Books and articles

Bakker M, Govers G, Jones R, Rounsevell M (2007), "The Effect of Soil Erosion in Europe's Crop Yields," Ecosystems 10
Borrelli P, Robinson D, Fleischer L, Lugato E, Ballabio C, Alewell C, Meusburger K, Modugno S, Schütt B, Ferro V, Bagarello V, Van Oost K, Montanarella L, Panagos P (2017), "An assessment of the Global Impact of 21st century land use change on soil erosion," Nat Commun 8
Panagos P, Borrelli P, Poesen J, Lugato E, Ballabio C, Alewell C, Meusburger K, Montanarella L (2015), "The new assessment of of soil loss by water erosion in Europe," Environmental Science & Policy 54

Chapter 8 – Drinking water

Databases

FAOSTAT: fao.org/nr/water/aquastat/data/query/results.html
OECD: data.oecd.org
OurWorldinData: ourworldindata.org
USGS: usgs.gov/centers/nmic/commodity-statistics-and-information

Reports from international organizations

EU (2018b), World Atlas of Desertification: wad.jrc.ec.europa.eu/download
UN (2019), The UN Water Development Report: unesdoc.unesco.org/ark:/48223/pf0000367306
FAO (2021), Progress on level of water stress, https://www.unwater.org/app/uploads/2021/08/SDG6_Indicator_Report_642_Progress-on-Level-of-Water-Stress_2021_ENGLISH_pages-1.pdf
UNESCO (2011), National Water Footprint Accounts: researchgate.net/publication/254859488_National_water_footprint_accounts_The_green_blue_and_grey_water_footprint_of_production_and_consumption
Water Footprint Network (2011), The Water Footprint Assessment Manual: waterfootprint.org/en/resources/publications/water-footprint-assessment-manual/

Books and Articles

Döll P (2008), "Wasser weltweit," Forschung Frankfurt 3/2008, https://www.forschung-frankfurt.uni-frankfurt.de/36050670/forschung-frankfurt-ausgabe-3-2008-wasser-weltweit-wie-gross-sind-die-globalen-susswasserressourcen-und-wie-nutzt-sie-der-mensch.pdf

Global Voices (2016), "Israel, eines der trockensten Länder der Welt, hat nun Wasser im Überfluss," https://de.globalvoices.org/2016/10/06/israel-eines-der-trockensten-lander-der-welt-hat-nun-wasser-im-uberfluss/
Hoekstra A und Mekonnen M (2012), "The Water Footprint of Humanity," PNAS 109 (9)
Lesch H (2016), "Die Menschheit schafft sich ab," Droemer Knaur, München

Chapter 9 – Energy

Databases

ESTAT: ec.europe.eu/eurostat/de/data/database
IEA: iea.org/data-and-statistics
OECD: data.oecd.org

Reports from international organizations

EU (2016), EU Reference Scenario – Trends to 2050: ec.europa.eu/energy/sites/ener/files/documents/20160713%20draft_publication_REF2016_v13.pdf
EU (2020), Global Energy and Climate Outlook 2019: Electrification for the low-carbon transition: publications.jrc.ec.europa.eu/repository/bitstream/JRC119619/kjna30053enn_geco2019.pdf
IEA (2019), World Energy Outlook 2019: iea.org/reports/world-energy-outlook-2019

Books and articles

BP (2019), BP Energy Outlook 2019, bp.com/content/dam/bp/business-sites/en/global/corporate/pdfs/energy-economics/energy-outlook/bp-energy-outlook-2019.pdf
McKinsey (2019), Global Energy Perspective 2019, mckinsey.com/industries/oil-and-gas/our-insights/global-energy-perspective-2019#

Chapter 10 – Raw materials

Databases

IAI: alucycle.world-aluminium.org/public-access
USGS: usgs.gov/centers/nmic/commodity-statistics-and-information
World Bank: data.worldbank.org

Reports and websites of international organizations

Copper Development Association (2017), Annual Data 2017: copper.org/resources/market_data/pdfs/annual_data.pdf
EU (2018c), World Atlas of Desertification: wad.jrc.ec.europa.eu/download
International Copper Study Group: icsg.org
International Copper Association (Stocks and Flows): copperalliance.org/about-copper/stocks-and-flows/
UNEP (2016), Global Material Flow and Resource Productivity: researchgate.net/publication/311101319_Global_Material_Flows_and_Resource_Productivity
World Steel Association: worldsteel.org

Books and articles

Bardi U (2013), "Der geplünderte Planet," oekom, München
Chen M and Graedel T (2016a), "A half-century of global phosphorus flows, stocks, production, consumption, recycling, and environmental impacts," Global Environmental Change 36
Dworak S and Fellner J (2021), „Steel Scrap Generation in the EU-28 since 1946 – Sources and composition", Resources, Conservation and Recycling, Vol.173, 2021
Fraunhofer (2016), "Technische, ökonomische, ökologische und gesellschaftliche Faktoren von Stahlschrott," Studie im Auftrag des BDSV, https://www.bdsv.org/fileadmin/service/publikationen/Studie_Fraunhofer_Umsicht.pdf
Glöser S, Soulier M, Espinoza L (2013), "Dynamic Analysis of Global Copper Flows," Environmental Science & Technology 47 (12)
Meadows D (1972b), "Die Grenzen des Wachstums," DVA, Stuttgart
Soulier M, Glöser S, Goldmann D, Espinoza L (2014), "Dynamic Analysis of European Copper Flows," Resources, Conservation and Recycling 129

Chapter 11 – The "ecological footprint"

Databases

Global Footprint Network: data.footprintnetwork.org

Books and articles

Randers J (2012b), "2052 – Der neue Bericht an den Club of Rome," oekom, München

Wackernagel M, Monfreda C, Moran D (2005), "National Footprint and Biocapacity Accounts 2005. The underlying calculation method," Land Use Policy 21

Chapter 12 – Species extinction and biodiversity

Databases

IUCN: iucnredlist.org/search
LPI: livingplanetindex.org/data_portal

Reports from international organizations

FAO (2018), The State of World's Fisheries and Aquaculture: fao.org/3/I9540EN/i9540en.pdf
IPBES (2019), Global Assessment Report on Biodiversity and Ecosystem Services: ipbes.net/sites/default/files/2020-02/ipbes_global_assessment_report_summary_for_policymakers_en.pdf
WWF (2020), Living Planet Report: https://www.wwf.de/living-planet-report

Books and articles

McRae L, Deinet S, Freeman R (2017), "The Diversity-Weighted Living Planet Index: Controlling for Taxonomic Bias in a Global Biodiversity Indicator," PLOS ONE 12 (1)
Urban MC (2015), "Accelerating extinction risk from climate change," Science 348

Chapter 13 – Forest loss – deforestation of rainforests

Databases

FAOSTAT: fao.org/faostat/en/#data
GFED: globalfiredata.org
GFW: globalforestwatch.org

INPE (brasilianische Behörde für Weltraumforschung): terrabrasilis.dpi.inpe.br/app/dashboard/deforestation
OurWorldinData: ourworldindata.org

Reports from international organizations.

FAO (2012c), State of the World's Forests: fao.org/3/i3010e/i3010e.pdf
FAO (2015b), Global Forest Resource Assessment 2015: fao.org/3/a-i4808e.pdf
FAO (2020), Global Forest Resource Assessment 2020: fao.org/3/ca9825en/ca9825en.pdf

Books and articles

Andela N, Morton D, Giglio L, Chen Y, van der Werf G, Kasibhatla P, DeFries R, Collatz G, Hantson S, Kloster S, Bachelet D, Forrest M, Lasslop, G, Li F, Mangeon S, Melton J, Yue C, Randerson J (2017), "A human driven decline in global burned area," Science 30
Van Der Werf G, Randerson J, Giglio L, Van Leeuwen T, Chen Y, Rogers B, Mu M, Van Marle M, Morton D, Collatz G, Yokelson R, Kasibhatla P (2017), "Global fire emissions estimates during 1997–2016," Earth System Science Data 9
Zimmer K (2019), "Erklärt: Der Amazonas produziert nicht 20% unseres Sauerstoffs," National Geographic 3/2019

Chapter 14 – Plastic waste in the oceans

Databases

OurWorldinData: ourworldindata.org

Reports and websites of international organizations

OECD (2022), Global Plastic Outlook: https://www.oecd-ilibrary.org/environment/global-plastics-outlook_de747aef-en
Heinrich-Böll-Stiftung (2020), Plastikatlas 2019: boell.de/sites/default/files/2020-02/Plastikatlas%202019%204.%20Auflage.pdf?dimension1=ds_plastic_atlas
Plastics Europe (v.a. Plastics – The facts 2019): plasticseurope.org/application/files/9715/7129/9584/FINAL_web_version_Plastics_the_facts2019_14102019.pdf
Plastic Insight: plasticsinsight.com
World Bank (2019), "What a Waste": openknowledge.worldbank.org/handle/10986/30317

Books and articles

Bai M, Zhu L, Peng G, Li D, An L (2018), "Estimation and prediction of plastic waste annual input into the sea from China," Acta Oceanologica Sinica 37

Bai M and Li D (2020) "Quantity of plastic waste input into the ocean from China based on a material flow analysis model", Anthropocene Coasts 3: 1-5]

Chassignet E, Xu X, Zavala-Romero O (2021), "Tracking Marine Litter with a Global Ocean Model: Where does it go? Where does it come from?", Front Mar. Sci. 23

[Geyer R, Jambeck J, Law K. (2017), "Production, use, and fate of all plastics ever made," Science Advances 3, 7

Jambeck J, Geyer R, Wilcox C, Siegler T, Perryman M, Andrady A, Narayan R, Law K (2015), "Plastic waste inputs from land into the ocean," Science 347

Lebreton L, van der Zwet J, Damsteeg JW, Slat B, Andrady A, Reisser J (2017), "River plastic emissions to the world's oceans," Nature Communications 8

Lebroton L und Andrady A (2019), "Future scenarios of global plastic waste generation and disposal," Palgrave Communications 5, 6

Lebroton L, Egger M, Slat B (2019), "A global mass budget for positively buoyant macroplastic debris in the ocean," Scientific Reports 9

Wu P, Xu R, Wang X, Schartup A, Luijendijk A, Zhang Y (2022), "Transport and Fate of all-time Released Plastics in the Global Ocean", preprint submitted to Earth Ar.Xiv.

Chapter 15 – Dead zones in the oceans – the P/N cycle

Databases

CGLS: land.copernicus.eu/global/products/lwq
FAOSTAT: fao.org/faostat/en/#data

Reports and websites of international organizations

EEA (2018), European Waters – Assessment of Status and Pressures 2018: eea.europa.eu/publications/state-of-water

OSPAR (2017), Eutrophication Status of the OSPAR Maritime Area: oap-cloudfront.ospar.org/media/filer_public/e4/85/e4858632-3d91-4245-8601-37792a2a987c/third_integrated_eutrophication_2017.pdf

Stockholm Resilience Center: https://www.stockholmresilience.org/research/planetary-boundaries/planetary-boundaries/about-the-research/the-nine-planetary-boundaries.html

Books and articles

Canfield D, Kristensen E, Thamdrup B (2005), "Aquatic Geomicrobiology," Advances in Marine Biology 48
Chen M und Graedel T (2016b), "A half-century of global phosphorus flows, stocks, production, consumption, recycling, and environmental impacts," Global Environmental Change 36
Diaz R und Rosenberg R (2008), "Spreading Dead Zones and Consequences for Marine Ecosystems," Science 321
Downing JA, Prairie YT, Cole JJ, Duarte CM, Tranvik LJ, Striegl RG, McDowell WH, Kortelainen P, Caraco NF, Melack JM, Middelburg JJ (2006), "The global abundance and size distribution of lakes, ponds, and impoundments," Limnology and Oceanography 51 (5)
Fowler D, Coyle M, Skiba U, Sutton M, Cape JN, Reis S, Sheppard L, Jenkins A, Grizzetti B, Galloway J, Vitousek P, Leach A, Bouwma, Butterbach-Bahl K, Dentener F, Stevenson D, Amann M, Voss M. (2013), "The global nitrogen cycle in the twenty-first century," Philos Trans R Soc Lond B Biol Sci. 368
Izmailova A und Rumyantsev V (2016), "Trophic status of the largest freshwater lakes in the world," Lakes and Resevoirs 21 (1)
Myanza O, Rwebugsia R, Mwinjaka O (2018), "Evidence of eutrophication in the Tanzania sector of Lake Victoria," African Journal of Tropical Hydrobiology ans Fisheries 16 (2)
Payton A und McLaughlin L (2007), "The Oceanic Phosphorus Cycle," Chemical Reviews 107,2
Steffen W, Richardson K, Rockström J, Cornell SE, Fetzer I, Bennett EM, Biggs R, Carpenter SR, de Vries W, de Wit CA, Folke C, Gerten D, Heinke J, Mace GM, Persson LM, Ramanathan V, Reyers B, Sörlin S (2015), "Planetary Boundaries: Guiding human development on a changing planet," Science 347
Voss M, Bange H, Dippner J, Middelburg J, Montoya J, Ward B (2013), "The marine nitrogen cycle: recent discoveries, uncertainties and the potential relevance of climate change," Philos Trans R Soc Lond B Biol Sci. 368

Chapter 16 – Pollutants in the environment

Databases

Bundesanstalt für Gewässerkunde: bafg.de
State of Global Air: stateofglobalair.org

World Air Quality Index: waqi.info/de/
OurWorldinData: ourworldindata.org

Reports of international organizations

EU (2004), Mercury Flows in Europe and the World: ec.europa.eu/environment/chemicals/mercury/pdf/report.pdf

EU (2017), Tackling Mercury Pollution in the EU and Worldwide: ec.europa.eu/environment/chemicals/mercury/pdf/tackling_mercury_pollution_EU_and_worldwide_IR15_en.pdf

Green Cross (2016), World's Worst Pollution Problems: studylib.net/doc/11193626/the-world%E2%80%99s-worst-pollution-problems--the-top-ten

UNEP (2018), Global Mercury Assessment: unenvironment.org/resources/publication/global-mercury-assessment-2018

UNEP (2019), State of the Global Environment: unenvironment.org/resources/assessment/part-state-global-environment

Health Effects Institute (2019), State of Global Air 2019: stateofglobalair.org/sites/default/files/soga_2019_report.pdf.

Books and articles

Carlsson P, Breivik K, Brorström-Lundén E, Cousins I, Christensen J, Grimalt J, Halsall C, Kallenborn R, Abass K, Lammel G, Munthe J, MacLeod M, Odland J, Pawlak J, Rautio A, Reiersen L, Schlabach M, Stemmler I, Wilson S, Wöhrnschimmel H (2018), "PCBs as sentinels for the elucidation of Arctic environmental change processes," Environmental Science and Pollution Research 25

Chen L, Zhang W, Zhang Y, Tong Y, Liu M, Wang M, Xie H, Wang X (2018), "Historical and future trends in global source-receptor relationships of mercury," Science of the Total Environment 610–611

Karagulian F, Belis C, Dora C, Prüss-Ustün A, Bonjour, S, Adair-Rohani, H, Amann, M (2015), "Contributions to cities' ambient particulate matter: a systematic review of local source contributions at local level," Atmospheric Environment 120

Miyazaki K, Eskes H, Sudo K, Folkert B, Bowman K, Kanaya Y (2016), "Decadal changes in global surface NO_x emissions from multi-constituent satellite data assimilation," Atmos. Chem. And Phys. 10.5194

Sharma B, Sanka O, Kalina J, Scheringer M (2019), "An overview of worldwide and regional time trends in total mercury levels in human blood and breast milk from 1966 to 2015 and their associations with health effects," Environment International 125

Zhong Q, Huang Y, Shen H, Chen Y, Chen H, Huang T, Zeng E, TaoZhong S (2017), "Global estimates of carbon monoxide emissions from 1960 to 2013," Environmental Science and Pollution Research 24

Further References

Databases

OHI: oceanhealthindex.org
UNEP: geodata.grid.unep.ch

Books and articles

Edenhofer O und Jakob M (2017), "Klimapolitik," Beck, München
Gore A (2014) "Die Zukunft – Sechs Kräfte, die unsere Welt verändern," Siedler, München
Horx M (2003), "Future Fitness," Eichborn, Frankfurt
Horx M (2013) "Zukunft wagen," DVA, München
Land K-H (2018), "Erde 5.0," FutureVisionPress, Köln
Pinzler P und Sentker (Hrsg.) (2019), "Wie geht es der Erde?," Komplett-Media, München
Ripple W, Wolf C, Newsome T, Galetti M, Alamgir M, Crist E, Mahmoud M, Laurance W (2017), "World Scientists' Warning to Humanity: A second Notice," BioScience 67
Smith L (2011), "Die Welt im Jahr 2050," DVA, München
Welzer H und Wiegandt (2011), "Perspektiven einer nachhaltigen Entwicklung," Fischer, Frankfurt